A CRISIS OF LOYALTIES

The Destruction and Abandonment of the Gosport Navy Yard

Stephen Chapin Kinnaman

Series in American History

VERNON PRESS

www.vernonpress.com

In the Americas:
Vernon Press
1000 N West Street, Suite 1200
Wilmington, Delaware, 19801
United States

In the rest of the world:
Vernon Press
C/Sancti Espiritu 17,
Malaga, 29006
Spain

Series in American History

Library of Congress Control Number: 2023949180

ISBN: 978-1-64889-808-2

Cover design by Vernon Press.

Images on the front cover: Destruction of the United States Navy-Yard at Norfolk, Virginia.

Image on the back cover: Plan of the Harbor of Norfolk and Portsmouth, 1861.

To my maternal grandmother,
Massie Moore Chapin,
an ardent Virginian.

TABLE OF CONTENTS

LIST OF ILLUSTRATIONS ix

INTRODUCTION xi

MAPS xv

CHAPTER 1
Gosport Navy Yard 1

CHAPTER 2
Charles Stewart McCauley 11

CHAPTER 3
The Coming Ordeal 21

CHAPTER 4
Old Dominion's Unwanted Decision 31

CHAPTER 5
Lincoln Takes Command 39

CHAPTER 6
Cumberland Arrives at Hampton Roads 49

CHAPTER 7
Virginia in the Balance 59

CHAPTER 8
Fort Sumter Surrenders 69

CHAPTER 9
Merrimack—The Prize 77

CHAPTER 10
Virginia's Agony, Wise's Insurrection 89

IMAGES 95

CHAPTER 11
Paulding's Flying Visit 119

CHAPTER 12
Virginia Votes for Secession 127

CHAPTER 13
The Commodore's Decision 137

CHAPTER 14
Alden's Shocking Report 147

CHAPTER 15
Norfolk Showdown 157

CHAPTER 16
McCauley's Scuttling 167

CHAPTER 17
The Die is Cast 177

CHAPTER 18
Paulding's Destruction 187

CHAPTER 19
The Aftermath 199

CHAPTER 20
The Accounting 211

APPENDIX 1
Annotated Calendar, March and April 1861 223

APPENDIX 2
List of Officers at the Gosport Navy Yard 225

APPENDIX 3
Secretary Welles' Orders of 16 April 1861 231

APPENDIX 4
Fate of Warships at the Gosport Navy Yard 235

APPENDIX 5
Destination of the Guns Seized from the Gosport Navy Yard 239

APPENDIX 6
The Potential of Conversions to a Steam Frigate 243

NOTES 247

BIBLIOGRAPHY 273

INDEX 281

LIST OF ILLUSTRATIONS

MAPS

No. 1 *A Plan of Portsmouth Harbor,* 1781 xix

No. 2 *Plan of Operations at Craney Island,* 1813 xx

No. 3 *Map of the City of Norfolk and the Town of Portsmouth,* 1851 xxi

No. 4 Gosport Navy Yard from *Map of the City of Norfolk and the Town of Portsmouth,* 1851 xxii

No. 5 *Plan of U.S. Navy Yard, Norfolk, Va.,* 1828 to 1860 xxiii

No. 6 *Panorama of the Seat of War, Bird's Eye View of Virginia, Maryland, Delaware and the District of Columbia,* 1861 xxiv

No. 7 *The Key to East Virginia,* 1861 xxv

No. 8 *Plan of the Harbor of Norfolk and Portsmouth,* 1861 xxvi

No. 9 *Military Reconnaissance, Dept. Va.,* 1862 xxvii

IMAGES

No. 1 *Commodore Charles Stewart McCauley* 101

No. 2 *The Navy-Yard at Norfolk, Virginia* 102

No. 3 *Entrance to the Gosport Navy Yard* 103

No. 4 *The Steam Frigate Merrimack at the Norfolk Navy Yard* 104

No. 5 *The State Capitol, Richmond, Va.* 105

No. 6 *Governor Henry A. Wise of Virginia* 106

No. 7 *Secretary of the Navy Gideon Welles* 107

No. 8 *Flag Officer Garrett Pendergrast* 108

No. 9 *U.S. Sloop Cumberland* 109

No. 10 *Engineer-in-Chief Benjamin Franklin Isherwood* 110

No. 11 *Captain James Alden* 111

No. 12 *Lieutenant Commander Thomas O. Selfridge, Jr.* 112

No. 13 *Sectional View of the U.S. Steam Frigate Merrimac* 113

No. 14 *Rear Admiral Hiram Paulding* 114

No. 15 *U.S. Steamer Pawnee* 115
No. 16 *U.S. Steam Tug Yankee* 116
No. 17 *The Virginia (Merrimac) in Dry-Dock* 117
No. 18 *Destruction of the United States Navy-Yard at Norfolk, Virginia* 118

INTRODUCTION

A good number of accounts have been written about the dramatic events surrounding the destruction and abandonment of the Gosport Navy Yard in the opening days of the American Civil War. Often employing its modern name, the Norfolk Navy Yard, these treatments have usually formed chapters of books on related subjects or have appeared as articles in periodical publications. One can also find books about the history of the Norfolk Navy Yard, the town of Portsmouth, Virginia, in which it was located, and plenty of titles on the duel of the ironclads, *Monitor* and *Merrimack.* But try and find a focused, book length treatment of this traumatic moment in U.S. Navy history and it will soon be discovered that the shelves are bare.

The lack of a definitive treatment leaves a gap in our understanding of the compelling stories behind the yard's loss, each full of colorful characters reacting to complex events, all so easily left out of shorter versions. The brevity of narrative has enforced a homogeneity of accountability for an outcome forged by many clashing, individual decisions. And it also avoids a full assessment of the lingering questions about just what happened at Gosport. For example, the largest and most valuable man-of-war in the yard, the steam frigate *Merrimack,* was stranded and ultimately burned to her waterline in the chaos that overwhelmed Gosport. Why was she not timely removed to a place of safety? The navy yard's commandant, Charles Stewart McCauley, made a last-minute decision to scuttle his most vulnerable ships. What motivated this drastic act and was it justified? Two senior U.S. Navy officers, Commodores Paulding and Pendergrast, were present at Gosport during the days leading up to the navy yard's abandonment and destruction. What roles did they play in its fate? Newly minted Secretary of the Navy Gideon Welles was sworn into office less than seven weeks before Gosport was put to the torch. What responsibility should he bear for the loss of the Navy's premier dockyard? And would any of these questions matter if *Merrimack*'s hull had been truly destroyed, and her reincarnation into the ironclad that battled *Monitor* had never happened?

The story of the loss of the Gosport Navy Yard has deep roots, embedded in the impending crisis that swept the United States and triggered South Carolina's secession and the creation of the Confederacy. When the fledgling Lincoln administration entered the scene in March 1861, the cracks in the Union were too large to paper over. New to their jobs, the president and his cabinet struggled to cope with rapidly unfolding events. Symptomatic of their internal turmoil was a pair of uncoordinated naval expeditions to relieve the besieged Forts Sumter and Pickens. The first failed miserably and when the action was over, Secretary of the Navy Gideon Welles had learned the lesson that the Navy's priorities took a second place to those of the Lincoln administration. Topping the list of losers was the Gosport Navy Yard.

Greatly complicating President Lincoln and Navy Secretary Welles' jobs was Virginia, the state that was home to the Gosport Navy Yard. It was also a slave state wracked by indecision over whether it should remain loyal to the Union or embrace its southern brethren and join the Confederacy. Each measure the federal government took to suppress the rebellion exerted its equal and opposite reaction on the State Convention delegates convened in Richmond to determine the Old Dominion's future. Stirring up the delegates' ire over Lincoln's every move was ex-Governor Henry Wise, a vocal advocate of his distinctive doctrine of "fighting in the Union." Wise's exhortations soon morphed into a personal insurrection aimed by his conspirators at federal installations in Virginia. The Gosport Navy Yard was one of their prime targets.

Meanwhile, at Gosport, Commodore McCauley presided over the nation's most valuable navy yard, a facility staffed overwhelmingly by officers hailing from Virginia. Such a preponderance of Southerners in a dockyard located within a large Virginian seaport was not considered unusual, but with sectional fissures tearing at the nation's fabric, these were not normal times. This Southern clique possessed a hierarchal power to sway every decision their commandant made, and in the days leading up to Virginia's secession, they invisibly played their hand to its fullest. Some called their behavior treasonous and there is merit to that argument. But before rendering judgment, put yourself in their shoes and consider the existential decision they faced as their states formally declared their severance from the old Union. The searing question for each of them was: *treason to whom*? To their flag? Or disloyalty to their families, neighbors and home state? Whatever choice they made carried with it some degree of betrayal.

Now back to the navy yard. With the principal background elements in place, the well-known version of events leading up to the fall of the Gosport Navy Yard can be traced. It begins with the old and feeble Commodore

McCauley and his indecision that resulted in the steam frigate *Merrimack*'s departure from Gosport being denied despite Engineer Isherwood's heroic efforts to repair her machinery. Then came news of Virginia's secession which sparked the mass exodus of nearly all of his officers. Meanwhile, Commodore Pendergrast, flag officer of the Navy's Home Squadron, sat idly by, avoiding active participation in the downwardly spiraling events. And soon enough McCauley's exaggerated fear of secessionist militia seizing the yard exerted its grip, triggering his ultimate act of desperation, the scuttling of his ships. Only hours later the arrival of Commodore Paulding and his Massachusetts Volunteers on the steamer *Pawnee* heralded relief. But they were too late and, regrettably, Paulding had no option other than to finish off the job McCauley had begun. The Navy's finest dockyard and over a thousand valuable naval guns were abandoned that night to the rebels, eleven of its warships put to the torch and the yard burned as Paulding's force escaped. And Commodore McCauley was judged by the court of public opinion to be responsible for the yard's loss.

But every one of these actions, many baffling and others confounding, flowed from a decision made by someone in charge based on what they knew at the time. A closer scrutiny of the motivations behind each of these decisions and actions has produced a clearer understanding of the events that culminated in the yard's destruction. Careful examination of the principal officers' backgrounds gave a deeper insight into what was in their minds during the hours leading up to the navy yard's burning. This combined focus has yielded a more nuanced explanation of McCauley's decision to hold back *Merrimack*, of Paulding's rush to burn the yard and run, and of opportunities for success missed by all three commodores present.

With any national disaster—and the loss of the Gosport Navy Yard in April 1861 was nothing less—it is mandatory to examine the roles of the responsible persons in charge. At the apex of leadership stood the Lincoln administration and more specifically, Secretary of the Navy Gideon Welles. The secretary has been awarded high marks for his efficient management of the Navy Department during the stressful years of the Civil War, and rightfully so. But in his first days on the job under a new administration struggling to find its way, he made mistakes. In the wake of the disaster at Norfolk, there is the appearance if not evidence that he discouraged any further inquiries into the loss of the Gosport Navy Yard, an event judged by a renowned naval historian as "one of the greatest blunders of the war."

Responsibilities and causes aside, the actions that clung to each day of the drama leading up to the burning of Gosport make a splendid introduction to the sectional conflict that followed. The participants' personalities, their life-wrenching decisions and the sheer historical imprint of what they did makes this story truly one to be savored by those who cherish Civil War history. Allow yourself to live in the world of Commodores McCauley, Pendergrast and Paulding and to experience their weight of command and be gripped by the emotional crossroads that entangled them all—a crisis of loyalties.

Many people encouraged me to write the story of the destruction and abandonment of the Gosport Navy Yard, none more so than my dear wife, Maureen Carroll. A welcome resource deserving praise are the *Norfolk Navy Yard* websites and *History Matters* blog managed by Marcus W. Robbins, historian and archivist. They helped immeasurably. And warm thanks to all of you who remain unnamed, especially those who generously gave their time to review the first draft of my manuscript. Your assistance and understanding made this book possible. Any errors of concept, fact or interpretation, however, are mine alone for which I accept full responsibility.

Stephen Chapin Kinnaman
Chappell Hill, Texas
September 2023

* * *

MAPS

NB: One regrettable feature of these historically interesting maps of Norfolk, Portsmouth, Gosport Navy Yard and the surrounding area is that nearly all of them disregard the standard convention of orienting their view with the north at the top of the image. This makes comparing one map with another more difficult than it should be, and for that, I can only caution the reader to be aware of the true north's bearing.

No. 1 *A Plan of Portsmouth Harbor*, 1781

The subtitle to this map is "Showing the works erected by the British forces for its defense," placing it in the context of the American Revolution. This plan illustrates the relation between Norfolk, Portsmouth and Gosport, the latter two separated by Crab Creek (salt meadow), echoing their English counterparts' similar geography. Mill Point is now Hospital Point; the modern naval hospital stands behind the battery. Only a pair of piers and a lone building hint at the location of the future Gosport Navy Yard. North is toward the bottom of the page (Digital Id http://hdl.loc.gov/loc.gmd/g3884p.ar145700 courtesy of the Library of Congress).

No. 2 *Plan of Operations at Craney Island,* 1813

This map shows the American defensive positions at the Battle of Craney Island, on 22 June 1813. A combination of regular and volunteer army forces bolstered by sailors from the U.S. frigate *Constellation* handily defeated a British attempt to capture Craney Island as a prelude to seizing the frigate. Midshipman Charles Stewart McCauley was among the island's defenders, where he gained a first-hand understanding of how a few determined men could choke movements up or down the Elizabeth River. North is toward the upper right corner (From *Lossing's Pictorial Field Book of the War of 1812,* Vol. 2, 679).

No. 3 *Map of the City of Norfolk and the Town of Portsmouth,* 1851

The growth of both Norfolk and Portsmouth seventy years after the Revolution is much in evidence, as is the now substantial footprint of the Gosport Navy Yard. The yard has been enlarged by the acquisition of land to the west but mainly to the south of Andrew Sprowle's original shipyard hard by Crab Creek. Pictures of the yard's main entrance (at Water and Lincoln Streets), the naval hospital's Greek revival façade and the receiving ship *Pennsylvania* are at the top right and center. North is to the left, lower corner (Digital Id http://hdl.loc.gov/loc.gmd/g3884n.ct003082 courtesy of the Library of Congress).

No. 4 Gosport Navy Yard from *Map of the City of Norfolk and the Town of Portsmouth,* 1851

This plan is an enlargement from Map No. 3, focusing on the Gosport Navy Yard. Note particularly the locations of the two enormous ship houses, A and B, at the yard's north end, and the granite-lined graving dock (far left Dry Dock, with entrance jetty walls) at the opposite side. The orderly, grid-like arrangement of the yard's many houses, shops and sheds is very evident. Officers' quarters, including the commandant's, stand along the yard's north Lincoln Street wall. The Elizabeth River's south branch provides ample sheltered water to moor warships in ordinary. Part of the St. Helena annex is visible on the bank opposite the ship houses (Detail from Map No. 3).

No. 5 *Plan of U.S. Navy Yard, Norfolk, Va.,* 1828 to 1860

Lull's Plate 2 published in 1874 represents the U.S. Navy's "approved plan" of the Gosport Navy Yard's grounds up to November 1860, three months after Commodore McCauley assumed the yard's stewardship. Little of substance has changed since 1851. Although not indicated in this plan, the massive sheer legs, or "sheers", used to hoist guns and other heavy weights in and out of ships stood on the edge of the wharf between Ship Houses A and B. North is to the left (From Edward P. Lull. *History of the United States Navy-Yard at Gosport, Virginia (near Norfolk).* Washington: Government Printing Office, 1874, Plate 2).

No. 6 *Panorama of the Seat of War, Bird's Eye View of Virginia, Maryland, Delaware and the District of Columbia,* 1861

The map was chosen to emphasize the connections through the Chesapeake Bay and along its tributaries between Norfolk and Washington, Baltimore and Richmond, with Fortress Monroe as their common nexus. The Bay Line's steamers tied all these points together with an integrated network of regular liner service. A fast steamer could cover the 145 miles between Baltimore and Fortress Monroe in under twelve hours. North is to the right (Digital Id https://hdl.loc.gov//loc.gmd/g3791s.cw0002000 courtesy of the Library of Congress).

No. 7 *The Key to East Virginia,* 1861

The focus of this map—the key to East Virginia—is the brooding presence of Fortress Monroe at Old Point Comfort. The fortress dominated the entrance to Hampton Roads and access to all points beyond including Richmond up the James and Norfolk on the Elizabeth. While the waters of Hampton Roads appear wide and inviting, they were fringed with numerous bars, banks and flats that complicated navigation. The Elizabeth River as it passes Craney Island provides a prime example: its narrow channel, marked by a light-boat, is bounded by Craney Island flat and Bush's Bluff shoal. Gosport Navy Yard is shown on the river's left bank just below the map's bottom border (Digital Id http://hdl.loc.gov/loc.gmd/g3880.cw0545000 courtesy of the Library of Congress).

No. 8 *Plan of the Harbor of Norfolk and Portsmouth,* 1861

This visually pleasing plan was published only a month before the destruction and abandonment of the Gosport Navy Yard. The broad expanse of the Elizabeth River's southern branch, punctuated by the "J" of St. Helena, is well displayed on this map. Numerous creeks indent the harbor's shoreline and the interconnection of Norfolk, Portsmouth and Gosport by water—the prime medium of nineteenth-century transportation—is obvious to the eye. Regional dominance of Norfolk's maritime infrastructure, if the numbers of named commercial piers are a reliable guide, is most striking. North is to the right (Digital Id http://hdl.loc.gov/loc.ndlpcoop/glva01.lva00014 courtesy of the Library of Congress).

No. 9 *Military Reconnaissance, Dept. Va.*, 1862

This map shows the results of a military survey of the area around Norfolk and the Gosport Navy Yard while still occupied by Confederate forces. The numerous batteries, virtually all populated by guns taken from the navy yard, are highlighted in red ink. It also illustrates the narrow the Elizabeth River ship channel leading into Norfolk and Portsmouth, running south past Sewell's Point and Craney Island. The three-fathom (18 ft) water depth contour graphically indicates how restrictive this waterway was for any deep-draft vessel (Digital Id http://hdl.loc.gov/loc.gmd/g3882h.cw0559000 courtesy of the Library of Congress).

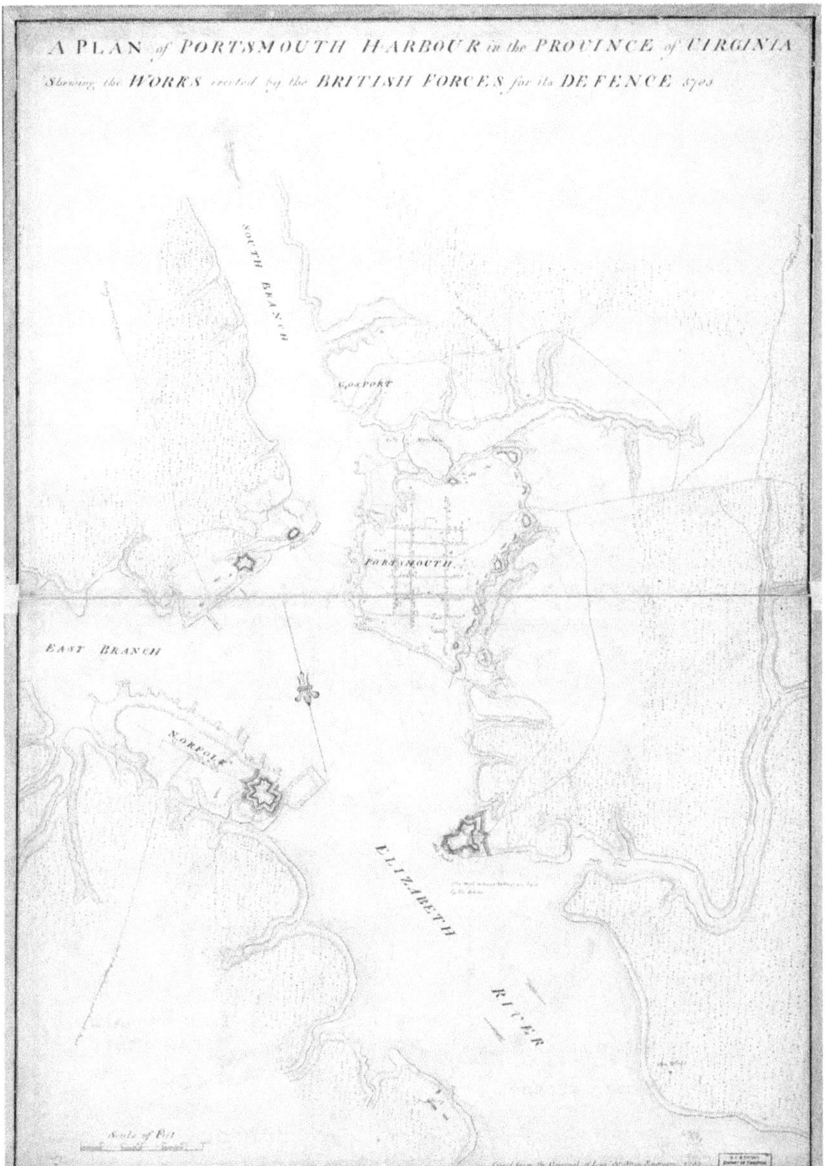

No. 1 *A Plan of Portsmouth Harbor,* 1781

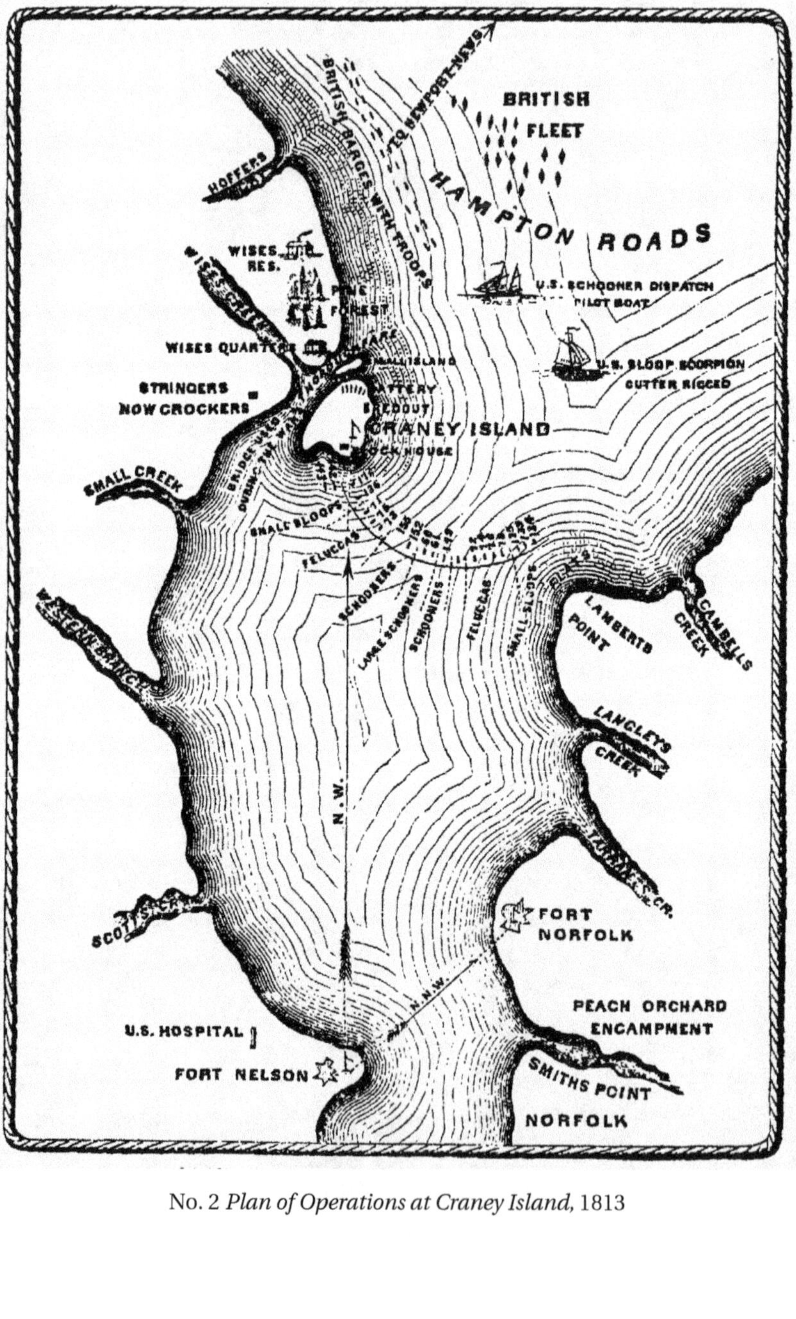

No. 2 *Plan of Operations at Craney Island,* 1813

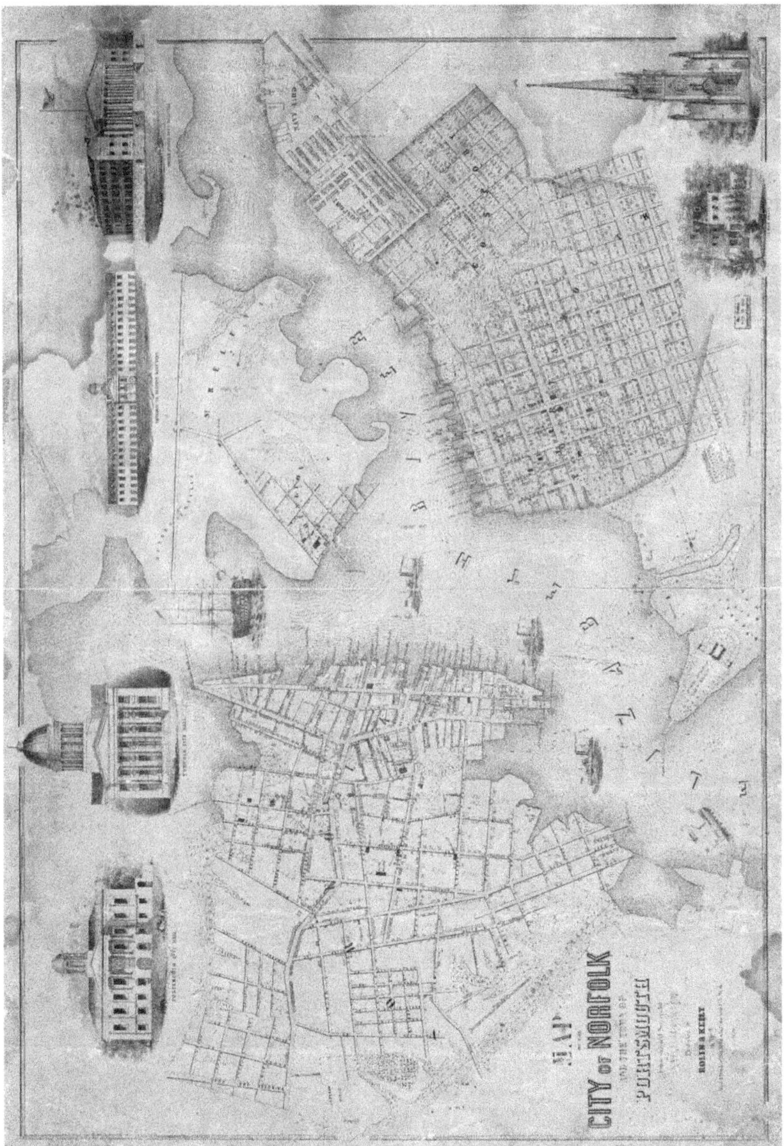

No. 3 *Map of the City of Norfolk and the Town of Portsmouth*, 1851

No. 4 Gosport Navy Yard from *Map of the City of Norfolk and the Town of Portsmouth*, 1851

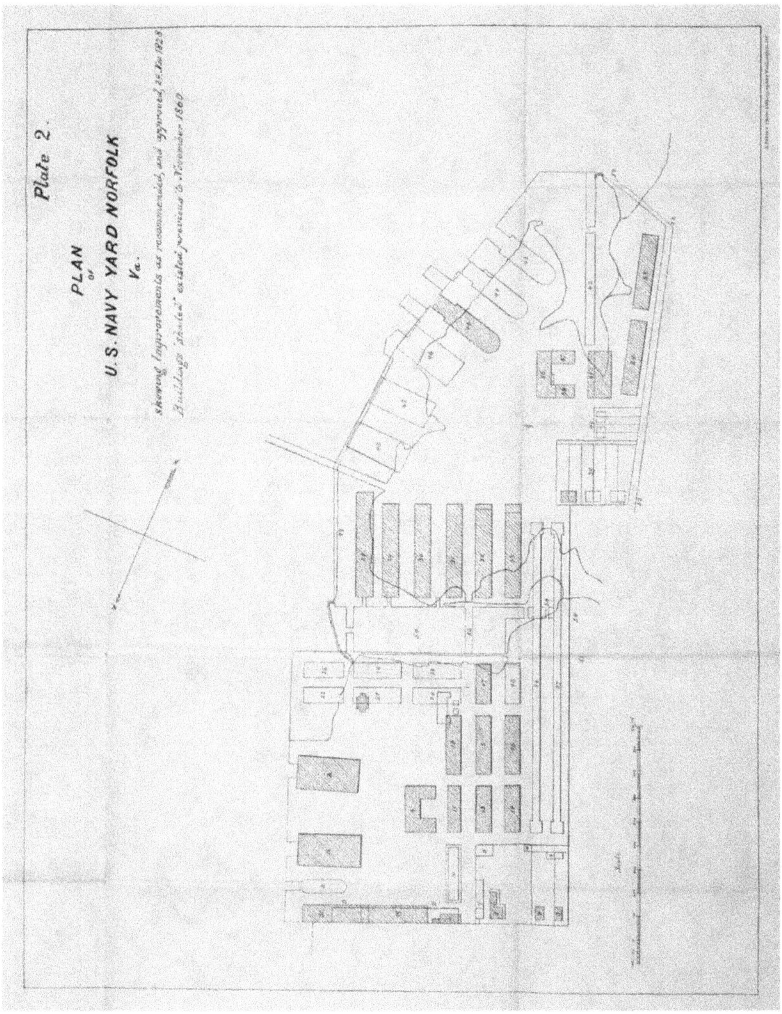

No. 5 *Plan of U.S. Navy Yard, Norfolk, Va.*, 1828 to 1860

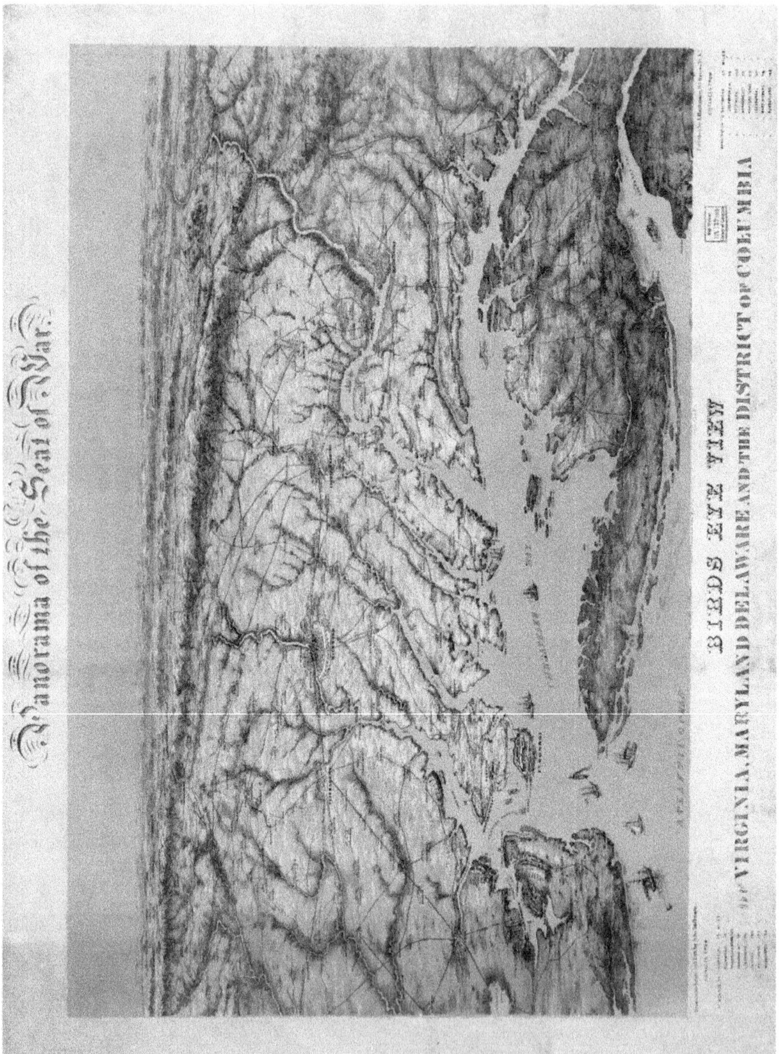

No. 6 *Panorama of the Seat of War, Bird's Eye View of Virginia, Maryland, Delaware and the District of Columbia*, 1861

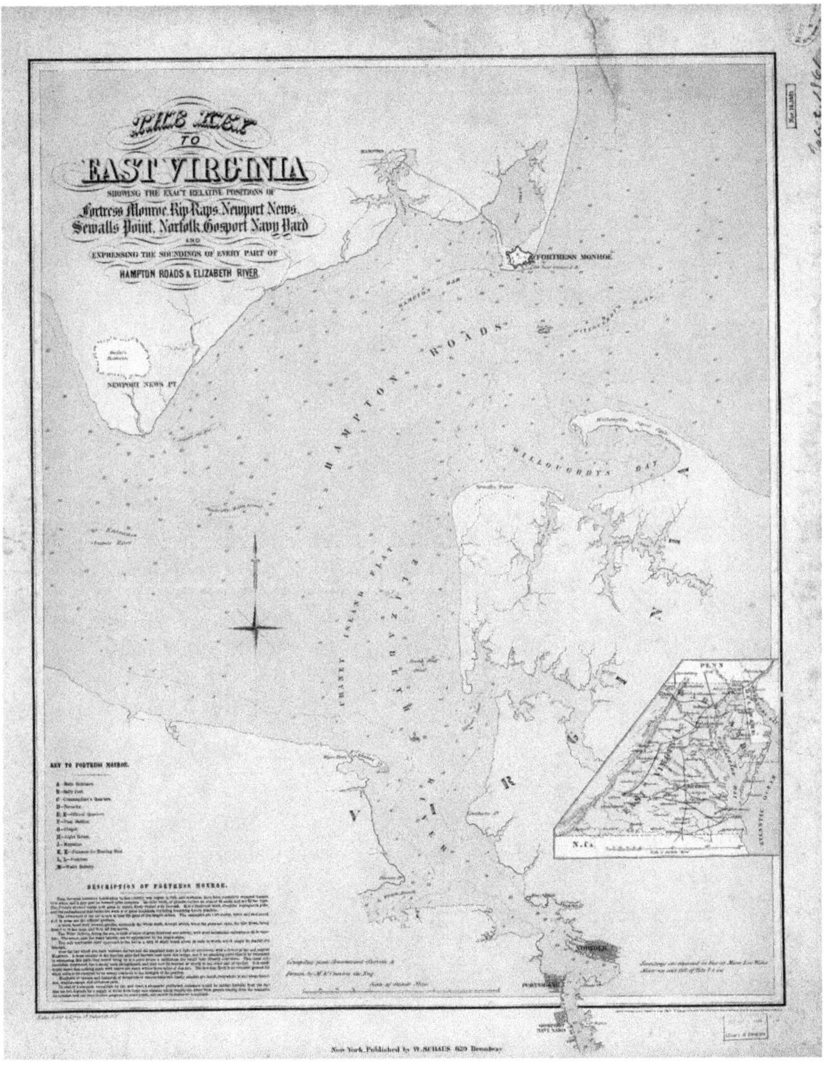

No. 7 *The Key to East Virginia,* 1861

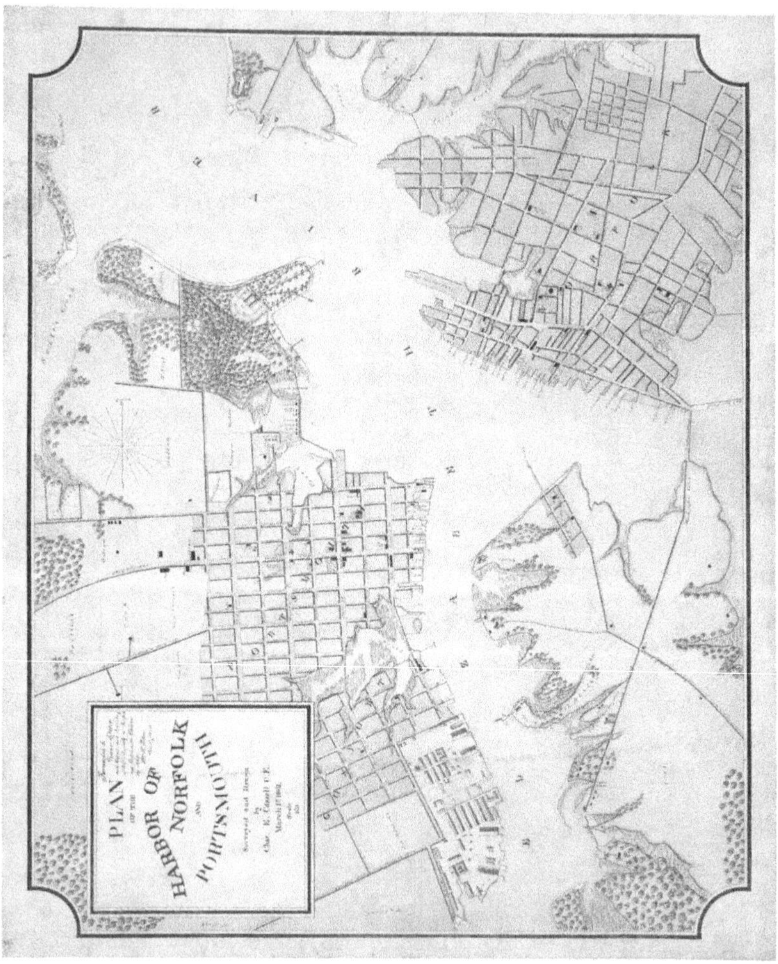

No. 8 *Plan of the Harbor of Norfolk and Portsmouth*, 1861

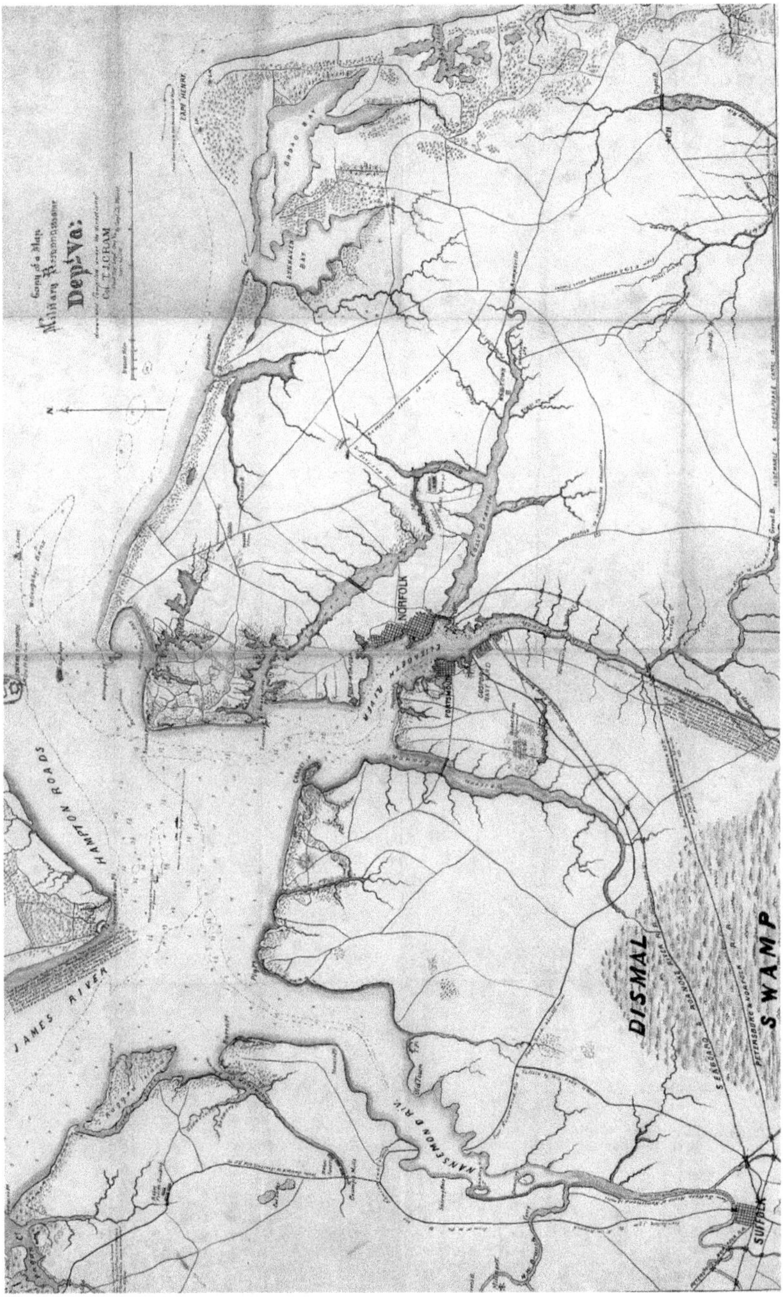

No. 9 *Military Reconnaissance, Dept. Va.*, 1862

CHAPTER 1

Gosport Navy Yard

On the 7th of August 1860, Captain Charles Stewart McCauley assumed command of the Gosport Navy Yard. Situated near Norfolk, Virginia, the dockyard was considered to be the United States' most extensive and valuable facility of its kind. It was certainly the nation's oldest, having been in service since before the American Revolution. Captain McCauley appeared to be a perfect choice for the yard's commandant, the honorific title assumed by the commander of a navy yard who was addressed as "Commodore." His record at sea had been exemplary, he could claim experience in managing a dockyard—McCauley had been the commandant of the prestigious Washington Navy Yard during the Mexican War—and at age sixty-seven he still possessed the vigor and mental activity sufficient for what was a largely ceremonial, shore-based job. Any questions about his fitness were easily brushed aside. After all, his eighty-two-year-old namesake uncle, Captain Charles Stewart of *Constitution* fame, was still in command of the Philadelphia Navy Yard.[1]

The navy yard over which Commodore McCauley presided rimmed the west bank of the southern branch of the Elizabeth River, and although styled Gosport, long before his arrival the land surrounding the yard had been incorporated into the adjacent town of Portsmouth. North across the Elizabeth River sprawled Norfolk, the biggest city in that part of Tidewater Virginia. Looking to the west, the Elizabeth flowed past the imposing Greek revival façade of the naval hospital and, just beyond the mud flats of Craney Island, emptied into the extensive waters of Hampton Roads. The capstone of this nexus of waterways was Old Point Comfort, less than fifteen miles by sea from Norfolk and forming the northern headland that defined the entrance to the "father of the waters," the Chesapeake Bay. Further to the east, outside of Capes Henry and Charles, loomed the vast expanse of the Atlantic Ocean. And 145 miles up the bay lay the vibrant entrepôt of Baltimore; nearly as far and in the same direction but on the Potomac sat the nation's capital, Washington, D.C.; and 75 miles up the James River, which emptied into Hampton Roads from the west, was Richmond, the capital of

Virginia.[2] Recognizing the strategic importance of America's largest estuary and its extensive body of inter-connecting rivers and streams, in the years following the War of 1812 the United States government erected a massive stone citadel at Old Point Comfort. Named Fortress Monroe, it also bore the monikers of "Gibraltar of the Chesapeake" and perhaps more apt, the "Key to the South."[3] Its purpose was to guard the bay's entrance and to ensure that the waters of Hampton Roads and the nation's most valuable dockyard remained secure from hostile foreign threats.

* * *

British colonists were keenly aware of the Elizabeth River's advantages, even without the cloak of protection later provided by Fortress Monroe. Advantageously situated at the mouth of a huge inland waterway system, it offered an ice-free, sheltered harbor for the paired towns of Portsmouth and Norfolk. Its southern branch beckoned as a prime location for ship repair with ready access to abundant stands of timber. In the early colonial days, the site at Gosport had been used for the careening of ships—heaving down their hulls to gain access to bottom planks. The numerous planters who inhabited the head waters of Virginia's spider web of creeks, streams and rivers were ready clients as were visiting ships of the Royal Navy. In the year 1767, an enterprising Scottish immigrant, Andrew Sprowle, purchased tracts of land south of Portsmouth's Crab Creek and established a shipyard at the careening location. Perhaps hoping to gain resonance from the Royal Navy's foremost dockyard at Portsmouth with its adjacent town of Gosport, Sprowle is thought to have chosen the name Gosport for his new facility. His strategy clearly worked; Sprowle was appointed as a Royal Navy agent in 1775.[4]

The American Revolution disrupted the lives of many loyal British subjects, none more so than Andrew Sprowle. Being branded a Tory, his valuable property was confiscated by the Commonwealth of Virginia and put to use maintaining its newly established navy. But the Elizabeth River was vulnerable to the wrath of the Royal Navy and after a brief occupation of Portsmouth and Norfolk, in May 1779 retreating British forces burned what was described in their correspondence as a "marine-yard [that] was the most considerable one in America" together with an incredible haul of 137 rebel vessels. It was the first time Gosport's shipyard would be put to the torch but, unfortunately, it would not be the last.[5]

Soon after the Revolutionary war ended, the land immediately adjacent to the shipyard was sold but the title to the yard itself was retained by Virginia, now one of the thirteen United States. It lay largely unused for the next ten years. Interest in Gosport surged in March of 1794 when Congress authorized the construction of six frigates for the U.S. Navy. The site was leased by the federal government for the construction of a 36-gun warship that would eventually bear the name of *Chesapeake*. Construction proceeded in a stop-start fashion, according to the perception of foreign threats, but Gosport managed to remain active fitting out privateers and supporting Commodore Truxtun's ships during the Quasi-War with France. It wasn't until May of 1800 that the frigate *Chesapeake* was finally commissioned. By this time the U.S. Navy had come to appreciate the value of the Gosport site. On 15 June 1801, a deed was executed which conveyed the shipyard's sixteen acres to federal ownership for $12,000, thus ending over twenty-five years of control by the Commonwealth of Virginia.[6] In hindsight, the U.S. government would probably have wished that a different wording had been used to frame a seemingly innocuous provision of the sale:

> That in case the Government of the United States shall at any time hereafter abandon the design of establishing a navy-yard at the place hereby ceded to the United States, or, after the establishment thereof, shall discontinue the same, then, and in that case, the property in the soil and the jurisdiction over the territory directed to be vested in the United States shall revert to this commonwealth, and shall be considered as the property and subject to the jurisdiction of the same, in like manner as if this act had never been made.[7]

* * *

The Gosport Navy Yard's first decade under U.S. Navy management saw the implementation of steady improvements to the site's facilities although, apart from the construction of several Jeffersonian gunboats, no vessels of significance were laid down. Mention of the brick wall that was to eventually surround the yard is first encountered in these years. During this period tensions were rising with Great Britain, fueled in part by the United States' efforts to assert its neutrality throughout the long years of the Napoleonic Wars. Further aggravating relations was the Royal Navy's habit of impressing American seamen. Matters came to a head on 22 June 1807 when Gosport's own *Chesapeake*, commanded by Virginia-born Commodore James Barron, sailed from Hampton Roads on what should have been a routine cruise. Just beyond the Virginia capes, she was confronted at gun-point by H.M.S. *Leopard*. Four sailors—presumably British deserters—were forcibly removed, one of whom was later hanged.[8] The national humiliation and rage that followed served to emphasize the need to more vigorously enforce the United States' sovereignty. At the same time, the incident further exposed the vulnerability of Norfolk, Portsmouth and the Gosport Navy Yard.

Apart from the occasional presence of a navy squadron at Hampton Roads or individual ships under repair at Gosport, no substantial improvements were made to area defenses. Twenty-one gunboats were afloat in the Elizabeth River by the end of the decade, but they were indifferently manned and no match for ocean-going warships. Declaration of war in 1812 brought the prospects of intensified threats to the Chesapeake Bay and its waters—the Royal Navy could be lurking anywhere just over the horizon. And so it was in early February 1813 that the U.S. frigate *Constellation*, under the command of Captain Charles Stewart, weighed anchor from Hampton Roads in an attempt to breakout into the Atlantic Ocean. But luck was not on his side. The pilots had warned him that a British squadron had just entered the Virginia capes. Sailing forth to investigate, Stewart realized he was now the quarry and quickly changed course for the nearest bolt hole—the Elizabeth River and the forts protecting Norfolk and Portsmouth. One of the officers serving on *Constellation* was Captain Stewart's nephew, Midshipman Charles Stewart McCauley, who would later become Gosport Navy Yard's commandant. In a quirk of fate, young McCauley celebrated his twentieth birthday on the same day the British force was sighted.[9]

Stewart feverishly prepared *Constellation* for the assault he knew would soon come. Anchoring her astride the river's most defensible location, the narrow channel between Craney Island and Tanner's Creek, Stewart sent down her upper masts and spars, strung up boarding nets and ordered her forecastle and quarterdeck carronades crammed with musket balls. He also coordinated his frigate's defense with Commodore John Cassin, the commandant of the Gosport Navy Yard and senior officer in command of the local fleet of gunboats. Cassin's small craft were moored adjacent to *Constellation* and the bigger ship's crew was distributed to create a "hedgehog" protective network. Meanwhile, batteries on nearby Craney Island were strengthened with heavy 18 and 24-pounders, and local militia deployed to thwart a British amphibious assault.[10]

Three times the Royal Navy attacked *Constellation*. Each assault was detected early enough to be handily thwarted. Finally, by early spring, the defenses of Craney Island and the Elizabeth River were complete and so after sinking four blockships in the main channel, *Constellation* was moved upriver and moored under the cover of Norfolk's protective batteries. In May the Navy Department decided to make better use of Captain Stewart; he was appointed to command the 44-gun frigate *Constitution* then fitting out in Boston. The still beleaguered *Constellation* passed into the hands of Master Commandant Joseph Tarbell; Midshipman McCauley remained attached to his ship.[11]

On 18 June 1813, Norfolk's defenders realized the balance of power in the Chesapeake Bay had shifted dramatically in favor of the British with the arrival of thirteen ships carrying 2,400 soldiers and marines complimented by two companies of "Independent Foreigners." And the scourge of the bay, Admiral George Cockburn, who would later mastermind the assault on and burning of Washington, D.C., was on the move. However, the Americans were not so easily intimidated. Spying an opportunity in the increased numbers of enemy ships in Hampton Roads, two days later acting Captain Tarbell launched an audacious pre-dawn gun-boat attack on the becalmed frigate *Junon.* Had not the wind picked up, she surely would have been an American prize. Midshipman McCauley participated in this widely admired action.[12]

Now that the hornet's nest had been stirred, retribution soon came to the stubborn defenders of Craney Island. Early on the morning of 22 June, British troops attempted a flank assault on the island from the mainland. No sooner had this wave been stopped by a fierce artillery barrage than an amphibious force of 1,500 sailors and marines filling fifty barges attacked the well-emplaced Americans. Hugging the shore in two columns to avoid fire from Cassin's mid-channel gunboats, the British inexorably closed until they came within range of Craney Island's defenders. *Constellation*'s officers and crew, who manned the island's two 24 and one 18-pounder, calmly eyed their approach. Out rang the order "Fire!" Their guns erupted in a deadly hail of rounds shot, canister and grape. On came the enemy—but then their lead boats grounded in the mud. The fierce cannonade smashed into the stalled barges, spilling their dead into the sea, forcing their retreat. British losses included five boats sunk or captured, six killed, twenty-four wounded and 114 missing of whom forty were taken prisoners by the Americans, who suffered no casualties.[13]

This successful joint operation by militia and naval forces ensured that the Elizabeth River, and by extension Norfolk, Portsmouth and the Gosport Navy Yard, continued free of enemy assault for the remainder of the War of 1812. *Constellation* remained bottled up at Norfolk for the duration but emerged unscathed. Midshipman Charles Stewart McCauley was an active participant in this signal American victory which has rarely received the recognition it deserves. One of the Battle of Craney Island's principal lessons was how the focused actions of a few determined men could paralyze the passage of the Elizabeth River. It was a lesson McCauley would not forget.

* * *

The end of the war brought a renewed appreciation of the urgent need to strengthen American sea power. Accordingly, following the passage of the landmark "Act for the Gradual Increase of the Navy of the United States," the keel of the 84-gun ship-of-the-line *Delaware* was laid down in August 1817 at the Gosport Navy Yard. The big ship was launched three years later and was finally commissioned in February 1828.[14] The need to enlarge Gosport's diminutive footprint was partially satisfied during the years 1826 to 1828 through the purchase of thirty-nine lots to the west of Portsmouth's Second Street and to the south of the existing yard along the banks of the Elizabeth. And perhaps realizing that the expanded yard required improved facilities to handle the Navy's biggest ships, construction of a graving dock was begun on the newly acquired land. This splendid granite-lined structure was opened on 17 June 1833, its initial customer fittingly being the Gosport-built *Delaware*. She was the first ship ever so dock in the western hemisphere.[15]

Another warship built in reaction to the humiliations of the late war, the 120-gun battleship *Pennsylvania*, was the world's largest vessel when launched in 1837 at the Philadelphia Navy Yard. She was simply too big and expensive to man during peacetime, so the Navy sailed her down the coast to Gosport where she was moored mid-stream and employed as a receiving ship. The forty acres of land across the Elizabeth beyond *Pennsylvania* were purchased nine years later and dubbed the Saint Helena annex. These grounds were used as a location for storage and repair of ordnance. In 1849 the War Department formally ceded to the Navy its property at Fort Norfolk, just downstream of and on the same side of the river as the city of Norfolk. The old fort was then employed as a magazine for the yard's gunpowder and shells.[16]

The steady pace of development during the antebellum period ensured Gosport Navy Yard's ship repair facilities were second to none. By 1860 the yard boasted of four building slips, two of which were permanently covered by ship houses; a granite graving dock; a timber or wet dock; a machine shop and a foundry; a saw mill with engine house; all manner of shops, lofts, sheds, store houses and offices; barracks and housing for officers; a gas lighting system; and a large timber pond. Adding to this was the nearby Portsmouth Naval Hospital annex, the Navy's largest such facility. The yard now covered 108 acres making it the nation's third largest in size behind Portsmouth (Kittery, Maine) and Brooklyn (New York). And by the end of 1860, seven war steamers had been built by its skilled shipwrights. Already prized because of its first-class physical plant and central location along the Atlantic seaboard, Gosport was accorded an extra degree of prominence due to its extensive ordnance depot which contained over a thousand guns. For many reasons the Gosport Navy Yard was considered to be America's premier naval dockyard.[17]

A few words at this point about the names Gosport, Portsmouth and Norfolk are offered in the hope of alleviating future confusion. The navy yard at the place called Gosport in Portsmouth, Virginia and across the Elizabeth River from the city of Norfolk had been, since its inception in 1801, formally named the Gosport Navy Yard. However, official Navy correspondence from this period almost always referred to it as "Norfolk." For example, McCauley's abstract of his Navy service shows that on 25 July 1860, he was ordered "To the Command of Navy Yard at Norfolk."[18] When describing events specific to the yard itself, Gosport or its full title, Gosport Navy Yard, will typically be employed. Happenings in the vicinity of the navy yard or relating to the area, especially if centered on the key city of Norfolk, will use the place name of Norfolk.

* * *

One characteristic of the Gosport Navy Yard that made it stand apart from its peers was the number of warships "in ordinary" afloat along the upper reaches of the Elizabeth River. "In ordinary" has a special meaning in the context of nineteenth-century navies. A ship "in ordinary" had been taken out of commission and laid up for the long term, either to be held in reserve for future mobilization or repair, or to await a decision as to whether to send her to the breakers. Such vessels were stripped of their masts, spars, sails and rigging, which were sent ashore for storage; their guns, shot and powder were hoisted out and placed in either the yard's gun park or magazine; and all stores, supplies and outfitting were removed. Typically, but not always, the upper deck of a ship in ordinary was roofed over to keep her hull dry inside. She was then securely moored in a location removed from maritime activity. It can be readily appreciated that once a vessel was placed in ordinary, she was incapable of operating as a ship-of-war until she was extensively refitted, revictualled, rearmed and brought to life with a full crew.

When Commodore McCauley took command of the navy yard in late 1860, Gosport was home to one ship under construction, five in ordinary, four under repair and its sole vessel in commission, the receiving ship *Pennsylvania*. Being built was the two-deck liner *New York*, whose frames had inhabited Ship House A for twenty years, an example of the Navy's penchant for minimizing its costs by keeping a ship on the stocks until required. The vessels under repair, which were decommissioned and substantially demobilized by the removal of their upper masts, rigging, sails, guns and supplies, were the new 40-gun steam frigate *Merrimack*—her machinery had been condemned and was due for replacement—two 22-gun sailing sloops-of-war *Germantown* and *Plymouth*, and the 4-gun brig-of-war *Dolphin*. Strung out in "rotten row" were five sailing vessels in ordinary, the ships-of-the-line *Delaware* and *Columbus*, and the frigates *Raritan*, *Columbia* and *United States*.[19]

* * *

Managing a navy yard like Gosport, with its comprehensive repair and building facilities, huge stocks of ordnance and large numbers of idle but valuable warships, necessitated a specially configured command structure. For understandable reasons, navies of this era tended to view anything requiring hands-on supervision to be a candidate for the same organizational system they applied to ships at sea.

To command the activities of the navy yard proper, a commandant was appointed as the senior officer, drawn from the list of the Navy's highest-ranking captains. Reporting to him was the yard's executive officer, typically a commander, who was supported by several lieutenants. Staff officers attached to the yard command structure were a surgeon and paymasters, the junior paymaster dubbed the inspector of provisions. Warrant officer positions of boatswain, gunner, carpenter and sailmaker completed the roster. The yard's wide range of ordnance functions were assigned to a senior officer, typically a commander in rank, supported by a lieutenant and a gunner. Because Gosport had a large naval hospital annexed to it, an additional surgeon and passed assistant surgeon were appointed for that service.

The yard's receiving ship, the enormous 120-gun *Pennsylvania*, required a parallel system of command: a commander as her captain, supported by several lieutenants, a surgeon, a passed assistant surgeon, a paymaster and a sailmaker. Since Gosport had no land-based saluting battery like other navy yards, it relied on its receiving ship's guns for that purpose—but no gunner was assigned to her. The yard's ordnance officers likely supervised the firing of *Pennsylvania*'s salutes.

Ships in ordinary were in the care of an officer specifically detailed for that purpose, a commander. The local "rendezvous" or recruiting station in Norfolk was managed by a pair of the yard's officers, a commander and a lieutenant. And Gosport was assigned a contingent of marines, under the watchful eye of a marine lieutenant colonel and his supporting first lieutenant. Their mission was to manage the yard's security and guard its premises and property, the receiving ship and all vessels in the yard's care.

Four civil officers completed the yard's command structure. Not commissioned officers, they were nonetheless hired according to their ability to meet exacting professional standards. They included a navy agent, a naval store-keeper, a naval constructor and a civil engineer.[20]

Just as Gosport was distinguished because of its large number of ships in ordinary and extensive ordnance facilities, so too was the pool of officers detailed to it—they were overwhelmingly from Southern, slave-holding states. A brief glance at the roster of its thirty-one commissioned and warrant officers (see Appendix 2) reveals that sixteen of them—more than half—were citizens of Virginia. A tilt toward Virginian officers serving in a navy yard located in their home state is not surprising, but the sheer weight of numbers stands out. If other Southern states are added, the bias becomes overwhelming. Two of the yard's officers hailed from North Carolina and five more from the slave states of Maryland, Delaware and the District of Columbia. Taken as a whole, nearly three-quarters of Gosport Navy Yard's officers were citizens of states south of the Mason-Dixon line.[21]

And not only were the overwhelming majority of the yard's officers Southerners, but many of them had put down roots in the area—they had either been raised in the vicinity, had established homes in Norfolk or were married to locals. Indeed, one officer fondly remembered that during the antebellum years "Norfolk was the Garden of Eden for naval officers" because so many had found their wives there.[22] The following examples paint the picture:

- Commander John R. Tucker, ordnance chief officer, was married to the former Virginia Webb, a Norfolk native, and had six children and a home in Norfolk.
- Gunner Charles B. Oliver, serving under Tucker, was married and living in Norfolk County.
- Passed Assistant Surgeon Dinwiddie B. Phillips of the naval hospital was married and resided in Portsmouth.
- Commander Arthur Sinclair, captain of the receiving ship *Pennsylvania*, was married to Norfolk native Lelia Imogen Dawley.
- Lieutenant Carter B. Poindexter, of the receiving ship *Pennsylvania*, was raised in Norfolk, married and had established his home there.
- Commander Richard L. Page, chief of the local rendezvous, was married to Norfolk native Alexina Taylor.
- Naval Constructor John L. Porter was born and bred in Portsmouth.[23]

In normal times this preponderance of officers with Southern inclinations might have sparked an interesting talking point, but in early 1861 the United States faced a sectional divide unlike anything that had come before. It was in those troubled days that the South's stranglehold on Gosport would ignite a true crisis of loyalties.

* * *

CHAPTER 2

Charles Stewart McCauley

Charles Stewart McCauley is something of an enigma amongst naval officers of his era. The perception that he played an ignominious role in the destruction and abandonment of the Gosport Navy Yard has caused historians to shun McCauley. No biography of him has been written, he left no extant memoirs and images of him are virtually non-existent. So what *is* known about Captain McCauley?

Charles Stewart McCauley was born on 3 February 1793 in Philadelphia, Pennsylvania, the eldest son of John and Sarah Stewart McCauley. His mother was the oldest sister of Captain Charles Stewart of *Constitution* fame, of whom her son was a namesake.[1] McCauley's parents were of Scotch-Irish origins and it seems a call to military service ran through their families; a grandfather on his Stewart side had been a colonel in the Royal Army. Close childhood friends of McCauley's uncle Charles included the naval notables of Stephen Decatur and Richard Somers. At that time, Philadelphia was the largest city in North America, the original thirteen colonies' undisputed center of culture and the capital of the newly established United States. It was a vibrant, exciting seaport for a young boy to grow up in.[2]

Official records of the U.S. Navy document that Charles S. McCauley was warranted as a midshipman on 16 January 1809, just shy of his sixteenth birthday.[3] However, other federal records show that on 15 May 1807, twenty months earlier, young McCauley had been issued a citizenship affidavit—a seamen's protection certificate—typical of those carried by American sailors to attest to their nationality.[4] The affidavit was issued in Philadelphia, witnessed by his father and gave his age as thirteen. Tying this together with abstracts of McCauley's career appearing in his obituaries stating that "he entered the navy at the age of 13 under Commodore Decatur ..." gives the firm impression that he went to sea with the Navy in some capacity prior to the date indicated by the official records.[5] Perhaps his uncle Charles had arranged for him to gain experience as a servant under the eye of his close friend, Captain Stephen Decatur, until such a time as he could be properly warranted. The irony of such a possibility is that Decatur was then "commanding naval forces in Norfolk," supervising the construction of fourteen Jeffersonian gunboats and later, captain of the 44-gun frigate *United States*. Decatur does not appear to have been the commandant of the Gosport Navy Yard but served as an active officer afloat until the eve of the War of 1812.[6] In this capacity, he would have had plenty of tasks to assign to his young charge, Charles Stewart McCauley. Small boats were the perfect school for teaching aspiring officer candidates the basics of command at sea.

* * *

As the War of 1812 tightened its grip across the United States, naval forces were deployed to best advantage. Charles McCauley, who had now been a proper midshipman for three years, was at some point attached to his uncle's ship, the 36-gun frigate *Constellation*, which began the war under refit at the Washington Navy Yard. The first chapter's narrative documented his active service on her in Virginian waters, at Hampton Roads and then Craney Island. By this time McCauley was a senior midshipman and may have, according to accounts, begun serving as an acting lieutenant.[7] This is confirmed by the Navy's documentation which records he was "app acting Lieutenant & ordered to Sloop of war Ontario at Baltimore 15 Sept 1813."[8] But *Ontario* was firmly blockaded into Baltimore by the same British squadrons that led the attacks on Craney Island, burned Washington, assaulted Fort McHenry and terrorized the Chesapeake Bay; she wouldn't leave until the war was over. So where did Acting Lieutenant McCauley go?

At this key period in McCauley's career, contemporary accounts tell that he "served under Commodore Chauncey as a midshipman."[9] Captain Isaac Chauncey was the commandant of the Brooklyn Navy Yard (New York) until May of 1813, and would have had little need for a midshipman to help run a dockyard. But soon after Chauncey was active on Lake Ontario. Late in the following year, he attempted to bring his British opponent, Captain Sir James Yeo, to bay before retreating into his stronghold, Sackets Harbor. The biographical dictionaries fill in the blank, saying that McCauley served as acting lieutenant on *Jefferson*, an 18-gun brig-of-war, on Lake Ontario in 1814.[10] The Navy's official records finally catch up with him, recording that on 9 December 1814 "at Sacketts Harbor promoted to Lieutenant."[11] At this remove it's impossible to say whether his advancement was well deserved but it was most certainly accelerated due to the press of war. Little more than five years after being warranted as a midshipman, Charles Stewart McCauley now held a lieutenant's commission in the U.S. Navy. It was his good fortune to have turned this vital corner in his career before the war ended. He was just twenty-one years old.

Meanwhile, Charles' younger brother, Daniel Smith McCauley, had at age sixteen answered the call to arms, obtaining a midshipman's warrant in February 1814.[12] His war would be thankfully short with peace being declared on Christmas Eve of that same year.

* * *

Hostilities may have ended but the Navy's squadrons beckoned. On 18 April 1815, Lieutenant McCauley was ordered to the new sloop *Erie* at Baltimore for what would turn into an extended deployment to the Mediterranean.[13] During the next five years, he served on board not only *Erie* but also her squadron mates, including the frigates *United States* and *Constellation.*[14] There is no record of McCauley taking any significant leave or a furlough before he was detailed for duty in May 1820 to undertake an even more far-ranging voyage, around Cape Horn and along the west coast of the Americas.[15] He joined the wardroom of the frigate *Constellation,* under the flag of Commodore Charles Ridgely, for what was the second regular deployment of a U.S. war vessel to the distant waters of the Pacific Ocean.[16] The stress of solo operations on such a remote station, far from support or resupply and constantly buffeted by the vicissitudes of the ongoing Chilean and Peruvian wars of liberation, would have placed extra burdens on the frigate's officers. By October of 1822, *Constellation* had returned to the east coast of the United States and McCauley was ordered to report at Philadelphia. Two months later he was furloughed for twelve months, and again in late 1823, for another twelve months.[17]

Perhaps because of a dearth of opportunity in the post-war Navy coupled with an offer that was too good to refuse, Charles Stewart McCauley traded his lieutenant's gold-trimmed uniform for the plain frock coat of a ship's master. For the next two years, he captained a merchant ship. McCauley's choice displayed a streak of initiative. It was not uncommon for naval officers of the day while "waiting orders" to fall back on a second profession: it served as a bolster to their income, kept their minds active and in McCauley's case, ensured he was abreast of the latest maritime developments. Where he sailed is not known but there is a tantalizing clue—during much of this same period his younger brother was also granted furloughs for travel "to India." McCauley must have been well regarded as a master for when he finally decided to return to the Navy, he turned down a handsome financial offer to stay with his commercial employer.[18]

McCauley's brother Daniel resigned from the Navy in April 1825, still a midshipman after eleven years of service.[19]

After a confusing shuffle of orders, in August of 1825, Lieutenant McCauley found himself temporarily detailed to the receiving ship at Boston.[20] On 28 December of the same year, he was sent the directive he must have expected—to report for duty under Captain William Crane on the sloop *Boston*.[21] Given his seniority, he was likely her executive officer, a demanding role for a lieutenant. Four years at sea on the Brazil Station followed, cruising the waters of the South Atlantic between Rio de Janeiro, Montevideo and Buenos Aires.[22] The next official reference to McCauley appears at the end of *Boston*'s cruise, when on 22 July 1829, Lieutenant McCauley was granted "leave unlimited."[23] He finally resurfaced a year later when on 28 May 1830 he was detailed to the command of the receiving ship at Baltimore. His charge was the diminutive, 51-ton gunboat *Fox*, a leftover from Porter's "Mosquito Fleet' used to suppress piracy in the West Indies.[24] But he *was* the captain. All things come to he who waits and so it was with McCauley—a year later he was promoted to the next rank up from lieutenant, master commandant (the modern equivalent of a commander). It had taken over sixteen years for him to make the jump. The date was 3 March 1831 and from then on McCauley proudly wore a single epaulet on his right shoulder.[25] Soon after the newly minted master commandant was placed in charge of Boston's recruiting rendezvous.

* * *

But Charles McCauley had his mind on more than just the Navy's business that year. On 25 October 1831, he married Lilia Elizabeth Dickins in a wedding that was described as a "gala occasion." The ceremony took place in Washington, D.C., the city of residence of his bride's parents. Charles McCauley was then thirty-eight years old and Lilia Dickins twenty-five, both rather ripe for such a move, especially the groom. Lilia's parents proudly hosted their eldest daughter's marriage and her eight siblings no doubt joined the celebration. McCauley's father John was then still active but his mother had passed away years before. The background of McCauley's in-laws provides a fascinating image of American society of that period. Especially compelling is the story of his father-in-law, Asbury Dickins.[26]

In the year of his daughter's marriage, Asbury Dickins was a civil servant, the chief clerk of the Treasury Department. He first earned his living in Philadelphia as a bookseller and publisher before sailing to England in 1803 to further his prospects. By October of that year, he had wed Lillias Arnot, the daughter of an estate owner from Balcormo, Scotland. But Dickins struggled to support his wife and growing family, eventually finding extra work as "chancellor" in the London office of the U.S. Consul.

There matters rested uneasily until 1812 when the outbreak of war upended his already tenuous security. A year later Dickins and his family left for Paris with the hope of promoting a light-amplifying "device" invented by the former naval officer, Lord Cochrane. Napoleonic France was rapidly collapsing but Dickins the survivor formed an enduring friendship with William H. Crawford, the U.S. minister to France. However, conditions worsened and the impoverished Dickins family was forced to quit France, sailing for America in October 1814. Years of meager living as a farmer followed. Dickins finally turned the corner in 1817 when he was appointed a clerk under Crawford, who was now at the Treasury. However, his travails were not over. Keeping up socially with his peers while maintaining a large family continually pressed on Dickins' financial resources. But as a well-placed bureaucrat, he had many friends in high places. Summing up his long career, Dickins' biographer wrote: "As a public servant for nearly half a century, Asbury Dickins probably knew more of the great and near great than any other man of his times."[27] What a perfect father-in-law for a U.S. Navy officer.

* * *

In January of 1832, Master Commandant McCauley was granted leave, extended twice, from his duties as commander of Boston's rendezvous. By mid-year, he had been transferred to Baltimore, again in charge of its rendezvous. He was now a father, his wife having given birth to a son who was named Charles Stewart. At the end of 1832, his big break finally came: an order to "hold yourself in readiness for sea service." On 12 February 1833, McCauley was appointed to the command of the sloop-of-war *Fairfield*, scheduled to depart for service with the Pacific Squadron. At age forty, McCauley had now been given his first independent command.[28] Although the pace of advancement in the antebellum Navy was sclerotic, he was on the move. Life was good.

What the fates give, they can take: on the last day of April, his nine-month-old namesake son died. Four days later, at his request, McCauley was detached from *Fairfield*.[29] His action demonstrated real courage—McCauley walked away from his first command. But he knew his wife needed him more than anything else, and so he followed his heart.

McCauley remained on leave of absence for over seventeen months when, on 22 October 1834, he again got a nod from the Navy. He was ordered to the command of the 18-gun sloop-of-war *St Louis*, then attached to the West Indies Station.[30] Informal accounts indicate that in the following year, McCauley commanded the squadron's sloop *Falmouth*. In August of 1835, he was granted three months' leave and spent all of 1836 waiting orders.[31] By this time, his father-in-law, Asbury Dickins, had moved into an influential position he would hold for the remainder of his working life, clerk of the Senate.[32]

In time McCauley's name came up for assignment to the Philadelphia Navy Yard as its executive officer. The date was 6 March 1837.[33] His orders couldn't have been a surprise; his uncle Charles Stewart was the yard's commandant. Four months after his arrival, McCauley had the opportunity to witness the launch of the world's largest warship, the 120-gun *Pennsylvania.* Before the end of the year, the behemoth was commissioned and under the command of his uncle, sailed around to the Gosport Navy Yard.[34] Five years later she began service as the yard's receiving ship. In time, McCauley and *Pennsylvania* would renew their acquaintance.

McCauley's family steadily grew during his stint at the Philadelphia Navy Yard. By the year of his appointment, he and his wife had already been blessed with a son, Asbury Dickins, and a daughter, Lilias Arnot. In late March of 1838 a second daughter, Maria Louise, was born. But by year's end, things began to go terribly wrong. The boy died in October at age five, of causes unknown. Then in June of 1839, both daughters died within two days of each other, probably from yellow fever. The eldest was three years old, the youngest just one.[35] In a span of nine months, McCauley buried three of his children. The specter of childhood death haunted parents of this era, but what Charles and Lilia McCauley had experienced was an unimaginable tragedy.

* * *

At the close of the final year of a troubled decade, Charles Stewart McCauley was promoted to the rank of captain. The date was 9 December 1839, just two months shy of his forty-seventh birthday. With his appointment, McCauley was again waiting for orders for his next Navy assignment. Given the slow turnover of peacetime openings for senior naval officers, fourteen months passed before the coveted order arrived. On 19 February 1841, McCauley was appointed to the command of the 84-gun ship-of-the-line *Delaware.*[36]

McCauley's new assignment placed him in the role of captain of a ship that served as a squadron flagship, in his case, of the Brazil Station under Commodore Charles Morris. He was in control of everything on his ship but had little scope to exercise independent command—McCauley was under the commodore's thumb. Morris' cruising ground was in the South Atlantic, off the coasts of Brazil, Uruguay and Argentina. In an unusual arrangement, in the middle of his deployment, Morris was ordered to change places with Commodore Charles Morgan of the Mediterranean

Station, thus moving *Delaware*'s area of operation into the wine-dark sea. Both cruising grounds would have been familiar to McCauley from his earlier service on *Boston* and *Erie*. What might have been a novel experience for *Delaware*'s captain was hosting royalty: in January 1842 he welcomed the emperor of Brazil, Dom Pedro II, on board and in September 1843, the king of Naples, Ferdinand II, was received with full ceremony.[37] *Delaware*'s cruise ended with her arrival at Hampton Roads, finally anchoring in Norfolk's harbor in March 1844. Captain McCauley was detached on 13 March and granted three month's leave.[38]

The returning sailor was welcomed by not only his wife Lilia but also by his two daughters—Rosa aged three and Julia, two. McCauley spent the remainder of 1844 and all of 1845 waiting orders, but at least he was granted time to catch up with his family. Early March of 1845 brought the birth of a son, who was named Charles in honor of the baby's father and late brother, the McCauley's first child.[39]

Early in 1846, McCauley's father died in Philadelphia at age seventy-nine. In mid-June of that year, McCauley briefly saw duty connected with the examination of midshipmen, a tedious task for most naval officers. But changes were in the wind. Triggered by disputes over Texas, in that same month, the United States went to war with Mexico. Captain McCauley was notified in late July that as of the 1st of September, he would be the Commandant of the Washington Navy Yard.[40] One of the nation's first dockyards, the Washington Navy Yard had been the Navy's crown jewel until it was burned by the British in 1813. Since then, it had been surpassed in importance by larger yards with easier access to the sea, such as Gosport. In the process, the Washington Navy Yard expanded its industrial expertise into the realm of ordnance technology. It was still a high-profile command for any naval officer, especially in times of war, one that was literally under the gaze of his superiors and visible from the Capitol building.

Five months after McCauley assumed command, a lieutenant named John Dahlgren reported to the yard for ordnance duty. He had been assigned to supervise the manufacture of rockets, but in time he expanded his remit to include a sweeping reform of how the Navy designed, manufactured and tested its guns. McCauley, who had never served as an ordnance officer, nonetheless recognized Dahlgren's genius and cooperated with him, giving enthusiastic support to Dahlgren's growing list of projects. By the end of McCauley's tenure at the Washington Navy Yard in October 1849, Dahlgren had introduced the first of a long line of his signature guns, a small bronze boat howitzer. More and larger cannons would follow.[41]

During McCauley's time at the navy yard, Lilia gave birth to a baby boy. Named Edward, he was their second living son and fourth child, all of whom were healthy and thriving.[42]

Following a relatively brief spell of waiting orders, Captain McCauley was ordered on 18 April 1850 to the command of the Pacific Squadron.[43] This took him into familiar waters from his cruise nearly three decades before as a lieutenant on *Constellation*. During his first two years as commodore, McCauley hoisted his pennant on the 44-gun frigate *Savannah*, shifting over to her sister *Raritan* for his final year. With the California gold rush in full swing, his squadron's prime responsibility was ensuring the safety of American mail steamers and U.S. citizens plying the Panama route. *Raritan* returned to Gosport in 1853 and was laid up in ordinary; on 21 January Captain McCauley was granted three months leave.[44]

* * *

A month after coming ashore, Charles Stewart McCauley celebrated his sixtieth birthday, no doubt surrounded by family and friends. During the remainder of that year, he was detailed to boards examining midshipmen and in 1854, to one that revised the Navy's "Table of Allowances." It would seem that McCauley's days afloat were behind him. But in April 1855 came an unanticipated order appointing him to the command of the Home Squadron. Long-festering animosity between the United States and Spain's captain general of Cuba had again reached the flash point. A year before the U.S. flagged mail steamer *Black Warrior* had been seized for charges of customs violations by authorities in Havana. Despite the resolution of this episode, harassment of American merchant ships continued, lately involving the steamers *El Dorado* and *Daniel Webster*. Spanish cruisers stopped and searched them, claiming a right of visitation. Commodore McCauley was directed to sail to Cuba in a muscular show of force designed to protect American interests.[45]

McCauley's mission was accomplished in a short span of two months. His conduct must have met the expectations of those in Washington who had sent him. "For his success ... he was publicly complimented on his return in June by President Pierce at a dinner at the White House."[46] Following his sudden rush of sea duty and presidential praise, Captain McCauley was appointed as a ranking member of the Navy's Efficiency Board.[47]

The impetus for the Efficiency Board, also known as the "Retiring Board," came from the large numbers of incompetent, incapacitated, or over-aged naval officers who clogged the Navy's peace-time roster, thus blocking the advancement of younger, more capable officers. Under an act of Congress passed that year, the president chose a board consisting of five captains, five commanders, and five lieutenants, with instructions to examine the records of all the Navy's officers and identify those who were incapable of rendering efficient service. McCauley together with Captains William Shubrick and Matthew Perry were the board's ranking officers. Within the relatively brief period of two months, the board identified 201 incapacitated officers. It recommended that forty-nine be dismissed, eighty-one retired on furlough pay and seventy-one retired on leave of absence pay. One of the "active list" captains moved to the "reserved list" was McCauley's uncle, Charles Stewart. The board's report created a huge uproar among those officers dismissed or retired. In time and after well-publicized protests, many were reinstated including Captain Charles Stewart.[48]

In the wake of McCauley's high-profile stint on the Efficiency Board, appointment to other boards followed, one for preparing a system of watch and station bills and another for recommending a code of signals subsequently adapted by the Navy. A spot of special service in 1858 enlivened the captain's daily routine and then he was again waiting for orders.[49]

In the year Captain Charles Stewart McCauley was ordered to the Gosport Navy Yard—1860—the federal census showed that he and his family were living in Washington, D.C. on F Street NW. His home was two blocks to the west of the Winder Building (still extant), which in turn sat across 17th Street from the old Navy Department building (long ago obliterated by the Eisenhower building). Five households away (using census terminology) at 1818 F Street stood the home of John Lenthall, the Navy's chief of the Bureau of Construction, Equipment and Repairs. It was a very convenient place to live for someone working at the Navy Department. Listed as members of the McCauley household were his wife Lilia (50), daughters Rosa (20) and Julia (18), and sons Charles (15) and Edward (12), plus cook Sarah McDermott (30), a native of Ireland. In terms of the value of estate owned, the McCauleys were better off than any other household appearing on the same census page, owning real estate worth $10,000.[50] No individual houses of that era now exist where they lived—the General Services Administration building covers the south side of that block—but their home might have looked something like the "F Street House" at No. 1925.

* * *

An overview of Charles Stewart McCauley's career on the eve of his assignment to Norfolk yields a picture of a well-rounded, sixty-seven-year-old naval officer. His role model uncle, Captain Stewart, was still active, a younger brother had served as a midshipman and his influential father-in-law remained in harness as clerk of the Senate. McCauley's human side had been fully displayed through his devotion to his wife and by fathering eight children, four of whose early deaths must have tested his every fiber. That he as a young man chose to take a break from his Navy career to captain merchant ships demonstrated initiative and an independent streak. As for other more nuanced elements of his character, McCauley's response to the crisis that enveloped the Gosport Navy Yard will serve to fill in the blanks.

McCauley's sea service was gained on five of the Navy's seven cruising grounds: the Mediterranean, Brazil, Pacific, West Indies and Home Stations. He held the independent command of a sloop-of-war, was captain of a ship-of-the-line, served as a squadron commodore, and undertook a high-profile special service command for which he was complimented by the president. The Navy took special care to record the sea service of each of its officers, using a year-month-day format. McCauley's record of 22-3-17 was exceeded by a few other officers. The next captain senior to him, Francis H. Gregory, who would be recalled from retirement at the outbreak of the Civil War and put in charge of the "Monitor Bureau," had time equaling only 19-10-17.[51]

As far as shore duty, McCauley had experience with all functions connected with a navy yard: command of a rendezvous and receiving ship, and as an executive officer and commandant. He sat on any number of boards, most notably one designed to weed out old, feeble officers.

Of the Navy's seventy-eight captains listed in the 1861 *Navy Register*, McCauley was ranked number seven. And like other Navy captains of his seniority, Charles Stewart McCauley could claim to have been blooded in combat against a worthy foe, the British. In McCauley's case, he gained the lion's share of his experience in Virginian waters, Hampton Roads and, most indelibly for a young midshipman, at the Battle of Craney Island.

CHAPTER 3

The Coming Ordeal

The year 1860 saw political upheaval as never before witnessed in American history. There was no escaping the impending crisis, only the final contours of its outcome were in doubt. The Republican party's presidential candidate, Abraham Lincoln, had expressed his belief that slavery must be placed "in the course of ultimate extinction." As for the Democratic party, it was hopelessly splintered and fielded three very different candidates. Election Day, 6 November 1860, saw two separate but simultaneous national elections: "Black Republican" Abraham Lincoln against Stephen Douglas in the free states and Bell versus Breckinridge in the slave states. When the ballots were tallied Lincoln had only thirty-nine percent of the popular vote, but because of the workings of the Electoral College system, he was declared the winner by a wide margin. Thus, the nation awoke to the news that it would soon be led by a president who had been opposed by sixty-one percent of the voters. America was now truly a house divided.[1]

South Carolina, long the champion of states' rights, quickly threw down the gauntlet. Following hasty approval by state legislators, a convention was opened in Charleston to address the burning issue of secession. Just three days later, on 20 December 1860, South Carolina's ordinance of secession was unanimously approved. For the first time in the Republic's history, a state had declared itself free of the federal Union—the very core of the United States. South Carolina's bold action inspired other Southern states to quickly follow suit. On 9 January 1861 Mississippi severed its ties with the Union and on the next day, 10 January, Florida followed in its footsteps. By the 1st of February, Alabama, Georgia, Louisiana and Texas had joined them. In the short span of just forty-two days, an irresistible secessionist movement had swept through seven Southern states. Like cracks in a sheet of ice, its fissures spread rapidly in all directions with unforeseen consequences.[2]

* * *

James Buchanan, the president of the United States, occupied an unenviable position. His office possessed the formal power to lead the nation during its sudden lurch toward rebellion but as a lame-duck, his ability to wield it diminished with each day. And while Buchanan opposed secession, he felt that the president didn't have the constitutional authority to suppress it. The crisis that was brewing was fundamentally different from previous political or domestic altercations—secession was an emotional popular reflex, immune to hasty legislative remedies. If the North's hostility toward slavery could be contained perhaps a compromise would have been within reach. But that was an impossible goal for any president.[3] So Buchanan's strategy largely consisted of temporizing until the president-elect assumed office on 4 March 1861. And Abraham Lincoln, who could have exerted real power, quietly bided his time in Springfield, Illinois. As an outsider to Washington's politics and intrigue, he had much to learn and little time to do so. But as was becoming painfully clear with each new day, a lot of mischief would be done during the long months before his inauguration.[4]

As state after state seceded, the mantle of federal authority was quickly cast aside. Throughout the South, the stars and stripes were struck from the grounds of United States mints, custom houses, post offices, courthouses, arsenals and any unguarded army or naval facility. Even floating federal assets such as light boats and revenue cutters were seized. Had a sitting president wished to wield the might of the United States armed forces, he would have been confronted by the fact of how meager it was. The posture of its military was that of a nation that had been at peace for more than a decade, its size dictated by a parsimonious Congress. The U.S. Army consisted of fewer than 17,000 troops and officers, the majority of which were garrisoning forts protecting the nation's vast western territories.[5] Naval forces were even weaker, with some 7,500 enlisted men and 1,300 officers manning its fifty warships in active commission. At best half of its vessels were steamers, with the greater part of its active-duty fleet forward deployed on the Navy's traditional foreign cruising grounds. Only the Home Squadron had assets available along the nation's eastern seaboard but, because of vexing problems with Mexico, many of them were stationed off the port of Vera Cruz.[6]

Further, if the commander-in-chief ordered the army or navy into action, what was the stated policy of the federal government underpinning his orders? Just what form would these orders take? What specific instructions would be issued? The officers of the U.S. Army and Navy had been trained to fight foreign foes. Leveling a rifle at one's fellow citizens would constitute a novel circumstance beyond their experience, one fraught with fearful consequences. With only their individual judgments as a guide, could the decision to spark a civil war be entrusted to them? The garrisons of forts and personnel manning navy yards were, to varying degrees, integrated into the local population, having families and friends living within sight of their barracks or dockyards. How was an officer in command of such a facility expected to react to demands by local militia, many of whom were neighbors, for the keys to his gate?

* * *

In the days immediately following South Carolina's secession the drama of rebellion played out first, predictably enough, at the forts commanding Charleston's harbor. Major Robert Anderson, whose federal army troops occupied the exposed Fort Moultrie, made a snap decision that would echo through history when, on the day after Christmas, he stealthily evacuated his entire command to the nearly impregnable, water-bound Fort Sumter. Overnight the fort's forbidding stone-faced battlements became a symbol of Union solidarity. The issue that had originally empowered the fire-eating secessionists, slavery, had now been usurped by an even more emotional cause, that of preserving the United States.[7]

Awaking to the dangers that an unchecked rebellion could wreak on federal authority, President Buchanan's secretary of the navy, Isaac Toucey, telegraphed a warning to his most vulnerable dockyard, the Pensacola Navy Yard. Sent on 3 January 1861, a week after Anderson's move, his telegram was addressed to the yard's commandant, Captain James Armstrong. "Be vigilant to protect the public property. The commanding officer at Fort Barrancas has been instructed to consult with you, and you will cooperate with him." The secretary's well-intentioned message would only be received by Armstrong six days later.[8] But Toucey's concerns about Pensacola must have persisted, for on the 5th he ordered the sloop-of-war *Macedonian*, then at the Portsmouth Navy Yard, to proceed immediately to Pensacola. And again, on the 9th, he instructed Captain Garrett Pendergrast, flag officer of the Home Squadron off Vera Cruz, to detach the 44-gun frigate *Sabine* and send her to Pensacola. In both cases, Toucey chose not to inform Armstrong of his dispatch of reinforcements.[9]

The newest of the Navy's shipyards, Pensacola had been established in 1826 on the north shore of a sheltered bay of that name on Florida's western panhandle. It was the only navy yard, apart from Gosport, situated in a Southern slave-holding state. Pensacola's development had been spurred by the Mexican War and the sudden demand created for its facilities. Even so, by 1861 it still lagged far behind its northern counterparts, possessing but one slipway and a floating drydock. Only two ships, both steam sloops, had been built at Pensacola. Guarding the navy yard proper, which was surrounded by the village of Warrington, were the Army's Fort Barrancas to the west, and at the harbor's entry channel, Forts McRee and Pickens. While Forts Barrancas and McRee were vulnerable to land attack, Fort Pickens, sitting on the west end of Santa Rosa Island, was protected by the barrier island's exposed approach over a flat, narrow sand bar and its surrounding waters.[10] Because of the dearth of military personnel, only Fort Barrancas was garrisoned. Company G, First Artillery, composed of some fifty men temporarily under the command of First Lieutenant Adam Slemmer, was the Army's sole force in the area.[11]

Captain James Armstrong, who had been detailed to the command of Pensacola on 13 September 1860, stood fourteenth amongst the Navy's captains and was within a few days of his sixty-seventh birthday.[12] It would be his first experience as commodore of a navy yard. In a curious blend, Armstrong had been born in a border state, Kentucky; appointed to the Navy from a deep South slave state, Mississippi; and was a citizen of the abolitionist stronghold of Massachusetts. His career largely paralleled Charles Stewart McCauley's, both being appointed midshipmen in 1809.[13]

The navy yard James Armstrong commanded was indefensible. Its sprawling grounds were surrounded by a brick wall just two feet high; its only artillery—a saluting battery composed largely of 18-pdr carronades—was mounted on rotten carriages with no trucks; and its compliment of marines totaled thirty-eight.[14] Of the yard's civilian employees, it was later remarked that "I do not think the commodore could have obtained the aid of single workman to aid in the defense of the yard."[15] And of Armstrong's team of officers, a number of them had openly expressed their secessionist leanings.[16] But in those tense first days of the new year, his biggest worry was less the defense of Pensacola than "the suffering condition of many of the mechanics and laborers employed in this navy yard, in consequence of the nonpayment of wages due them since 1st of November last, the paymaster not having funds to pay the rolls."[17] The fissures of secession had spread far and wide. With rumors swirling in the air about armed insurrectionists on the march, Commodore Armstrong would need to quickly change his focus.

* * *

Back in Washington, D.C., President Buchanan crafted a surprisingly hawkish response to South Carolina's demands that Major Anderson evacuate Fort Sumter, declaring his intention to defend the fort "against hostile attacks, from whatever quarter they may come." But those strong words demanded action and so an expedition to reinforce Fort Sumter's garrison was quickly organized. On 5 January 1861, the unarmed mail steamer *Star of the West* sailed from New York, carrying some 200 troops and badly needed supplies. In the meantime, realizing his hasty move might provoke a war, President Buchanan was assailed by doubt and attempted to recall her—but the bird had flown. And so, four days later *Star of the West* boldly stood in Charleston's harbor. Greeted by hostile fire from shore batteries and receiving no support from Fort Sumter's guns, she promptly turned back. Buchanan's halting and uncoordinated attempt to relieve Fort Sumter garnered the first shots of the war. The United States flag had been fired on and Northerners were humiliated.[18]

As if to reinforce the Buchanan administration's inept performance, Navy Captain Samuel Barron was ordered to Norfolk on 7 January—two days after *Star of the West* had sailed—in a vain attempt to intercept her before she reached Charleston. His mission was to deliver dispatches to Captain Walker of the steam sloop *Brooklyn*, thought to be lying at Hampton Roads, ordering him to sea in pursuit of *Star of the West*. Note that by this date it was deemed necessary to hand-carry confidential orders to ensure their secrecy. But Barron found *Brooklyn* still at the Gosport Navy Yard.[19] Four days later he related details of his excursion to Captain Samuel Du Pont, then serving as commandant of the Philadelphia Navy Yard and one of the Navy's rising stars. Du Pont later recorded the following account of their conversation:

> Captain Barron came on Friday night ... he had been sent to Norfolk to communicate with the commander of the *Brooklyn*. He represents that place as disloyal, and twelve hundred workmen in the Navy yard are wearing a blue cockade [the badge of secession], ready to *take* the yard if "coercion" be attempted; the commandant there is the weakest of the weak. ... The affair of the *Star of the West* was very bunglingly managed.[20]

Striking are the observations relayed by Barron about Gosport's visible workforce disloyalty and Du Pont's harsh judgments on Commodore McCauley's character and the *Star of the West* expedition. Future events would lend credibility to the claim of workforce disloyalty; the charge of McCauley being weak should be borne in mind during evaluations of the crisis that overtook the Gosport Navy Yard. As for Barron and his mission to *Brooklyn*, she didn't sail until the 9th, the same day the door was firmly slammed in *Star of the West*'s face, making her mobilization one more exercise in futility.[21] Open war had been avoided, but for how much longer?

* * *

The pace of events suddenly began to pick up in Pensacola. On the day before Florida formally announced its secession, Lieutenant Adam Slemmer received orders from Army headquarters to defend either of the two forts, Barrancas or Pickens. He quickly paid a visit to Commodore Armstrong to determine a joint plan of action. It was the same day Armstrong had belatedly received his instructions from Toucey dated the 3rd. They mutually "agreed … Fort Pickens was the only place that could be defended by the united forces of the Army and the Navy," and to that end, Armstrong promised to furnish the Army with what men and provisions he could spare. Further, he gave instructions to the commander of the 5-gun third-class steamer *Wyandotte*, then at the yard preparing for sea, to assist with towing a lighter and small boats across the bay to facilitate the relocation of Slemmer's company. A second Navy vessel, the sailing store ship *Supply*, was moored off Fort Barrancas to further support the Army's transfer. Slemmer's move was substantially completed on the morning of 10 January, the day Florida seceded.[22] In the meantime, Armstrong had increased the marine guard and security protocols at the navy yard's main gate and had issued instructions to the captains of both *Wyandotte* and *Supply* that in the event of the yard's capture, they were to fire only in their self-defense and if the yard was taken, to immediately seek safety and return to the station from which they had been originally ordered by the Navy.

The joint operation to garrison Fort Pickens was completed none too soon. On Saturday morning, 12 January, seven companies of well-armed and uniformed Florida and Alabama militia marched up to the navy yard's east gate. Backed up by 350 very visible troops, a Florida state commissioner demanded immediate surrender of the yard. Earlier that morning, Armstrong had ordered his marines out, armed and ready for trouble from insurrectionists. But this was something bigger, much bigger. After angst-filled moments during which he hurriedly destroyed his signal books, Armstrong reluctantly lowered his flag. It was a bow to the inevitable; he and his Army opposite had already determined Fort Pickens was Pensacola's only defensible position. As the commodore later explained in his report describing his actions:

> I had no specific instructions to guide me. I was not advised of the intended policy of the Government in dealing with these troubles, and wholly destitute of the necessary means for coping with so superior a force, I was left to the conclusions of my own judgement. … Could I dare assume so grave a responsibility as striking a blow, useless in practical results, and a signal for inaugurating all the dread horrors of a civil war? My impression was that conciliation and a forbearing treatment was the policy, apparently, of the Government.[23]

Wyandotte and *Supply* were not molested by the militia and got to sea, later returning to Fort Pickens under a flag of truce. There Lieutenant Slemmer was soon confronted by secessionist officers demanding his surrender. He managed to put off responding until the 16th when he gave his final answer, a blunt refusal. Together with Fort Sumter, a second hot spot had now been added to the cauldron of rebellion.[24]

* * *

James Armstrong was soon released under parole and made his way north to Washington, D.C., where on 25 January 1861 he submitted his report to Secretary of the Navy Isaac Toucey describing his surrender of the Pensacola Navy Yard.[25] On the following day, Armstrong again wrote to Toucey to "request that a court of enquiry may be instituted to enquire into all the facts and circumstances of said capture …"[26] The matter of the loss of a navy yard had now officially pressed itself into the secretary's domain, begging for a response. In the two weeks that had passed since the traumatic event, Pensacola must have been a subject of discussion, both formally and informally, within the Buchanan administration. But the horizon for Toucey's tenure in office was drawing closer with each day; his apparent lack of celerity seems to indicate a growing disengagement with public affairs.

That Secretary Toucey still harbored concerns over the safety of his one remaining navy yard in a Southern state—Gosport in Virginia—is demonstrated by a revealing order he issued the day *before* Captain Armstrong submitted his report. On the surface, his action might appear unconnected to what happened at Pensacola, but viewed as a whole, the events surrounding Toucey during those three days in late January appear to be joined by a common thread—the security of navy yards.

So first, consider Toucey's action. On 24 January Secretary Toucey issued the following written order to Navy Captain Levin Powell. "Proceed to Norfolk, Va., and after carrying out the verbal orders you have received from the Department return to your residence at Washington City."[27] Note the recourse to verbal orders and a trusted officer to carry out their intent. But what were Powell's orders? As no record of them or Powell's report back to the secretary exists, Toucey's later testimony at the Senate select committee hearings into the loss of the Gosport Navy Yard provides some of the missing information:

> I had no doubt about the safety of the yard. I took the precaution to send a very capable and faithful officer of the navy, Captain Powell, upon a secret mission to the yard, to confer with the veteran officer, Commodore McCauley, who was in command there and see that

> nothing was wanting to secure its safety. His report was entirely satisfactory to me.[28]

In subsequent testimony, Toucey admitted he was "unable to recollect" any details of Powell's visit, and as to whether he had received further requests for men or means to protect the yard, he replied, "not to my recollection." That he had gone through the motions of checking up on McCauley was, apparently, quite enough.

Captain Powell was not asked to testify at the hearings, but McCauley's recollection of his visit, given at the same forum, filled in the blanks:

> Towards the last of January ... 1861, Captain L. M. Powell, of the navy, came down to Norfolk, as a special messenger from the Navy Department, to confer with me in reference to the condition of the yard as far as its defences were concerned. I told him I did not fear anything in the shape of a mob, but that if a large body of organized troops should attack the yard my force was very inadequate to its protection, and urged reinforcements, which could be sent without exciting any suspicion or alarm; for, as the "Germantown" was then preparing for sea, and the "Plymouth," ... was also preparing ... , it would have been a sufficient pretext to send the officers, seamen, and marines for these two ships. He stated that I would receive orders for commissioning the "Plymouth" and "Dolphin," but he returned to Washington, and I never heard anything in relation to the matter.[29]

A search of the *Official Records of the Union and Confederate Navies in the War of the Rebellion* (*Official Records*) supports McCauley's statement that Toucey failed to follow through. Orders relating to the manning of *Germantown* and *Plymouth* were subsequently issued, but only after Lincoln's secretary of the navy, Gideon Welles, assumed office in March.[30]

Now, back to Captain Armstrong and his request for an inquiry into his conduct. It is important to understand that what Armstrong asked for, a court of inquiry, was a fact-finding body rather than the more formal general court-martial, which if convened, would try and assess guilt for serious infractions of the Articles of War.[31] Toucey agreed to Armstrong's request and on 2 February, he appointed three Navy captains and a civilian judge-advocate—the government's prosecutor—to enquire into the circumstances surrounding the capture of Pensacola. One of the naval officers appointed to sit on Armstrong's court of inquiry was Captain Levin Powell, who had recently been sent to check on McCauley at Gosport.[32] His selection probably had more to do with availability than any calculated motive.

Armstrong's court of inquiry was convened on 8 February in Washington, D.C., and continued in session until 19 February, its proceedings filling twenty-eight dense pages of the *Official Records.* The court's final opinion was read by its president on 20 February. It determined that "Commodore Armstrong was culpable in not taking any precautionary measures for the defense of his post and the public property thereat." The court also thought that he "should have induced a resistance ... against the forces brought against him" and that if he could not defend the yard, he should have removed all his men and public property from it and destroyed what could not be removed.[33] The findings of the court were then submitted to Secretary Toucey for a decision as to whether to further pursue the matter or let it rest. Perhaps because his term of office was due to end in less than two weeks, the secretary chose to do nothing. However, the court's findings must have been of enormous interest to all U.S. Navy officers, none more so than Commodore Charles Stewart McCauley.

* * *

While Secretary Toucey's Navy Department imperfectly coped with the loss of one of its dockyards, the forces of rebellion pressed forward with relentless energy. On the morning of Monday the 4th of February, delegates from six of the seven states that had seceded from the Union convened in Montgomery, Alabama's capital, to establish a provisional government. The birthing of the Confederate States of America raced ahead with undisguised enthusiasm, still riding the momentum that had propelled the first wave of secession. By 9 February—five days' later—they had adopted a provisional constitution and elected a president and vice president. On 18 February Jefferson Davis of Mississippi and Alexander Stephens of Georgia were formally inaugurated into their respective offices. With the former convention delegates now serving as a provisional legislature and its chief executives in place, the newest republic in the Americas had a fully functioning government.[34]

That the Confederacy was composed of only seven of the Union's fifteen slave states should have given members of the new government pause for reflection. Their cause would have been immeasurably stronger by the addition of those undecided border slave states. But their audacious achievement in such a short time—inconceivable in the months leading up to November's presidential election—was plenty enough excuse to celebrate. There was still time to win new converts. After all, Abraham Lincoln and his fellow Black Republicans wouldn't assume office for another two weeks. It was the dawn of Southern independence and the sun had yet to rise.

CHAPTER 4

Old Dominion's Unwanted Decision

That fateful Monday, the 4th of February, kicked off more than just the labor that gave birth to the Confederacy. It also witnessed the citizens of the Old Dominion turning out to vote for delegates to a state convention that would consider Virginia's stance within a visibly fractured Republic. At Willard's Hotel in nearby Washington, D.C., that same day saw the seating of delegates to a hastily organized Peace Conference convened in an attempt to adjudicate the nation's sectional differences. Chairing the Washington gathering was the Virginian and former president, John Tyler.[1]

The state of Virginia occupied an outsize position in the rapidly escalating disintegration of the Union. The first successful English settlement in North America, Jamestown, had been planted on its soil in 1607. Four of the United States' first five presidents hailed from the Old Dominion, including the revered "father of his country," George Washington; the author of the Declaration of Independence, Thomas Jefferson; and the "father of the Constitution," James Madison. Although Virginia's commercial progress had not kept pace with its illustrious past, partly due to the economic drag of slave-holding, the state nonetheless commanded considerable national respect.[2]

Virginia was also home to three prominent federal military installations of interest to all players in a potential conflict: a major arsenal at Harpers Ferry, Fortress Monroe commanding the Chesapeake Bay at Old Point Comfort and the strategically located and comprehensively equipped navy yard at Norfolk. One look at a map made several other points abundantly clear. As Virginians already knew, their position as a border state adjacent to the Northern free states greatly elevated their fugitive slave problems if compared to planters in the deep South. But even worse, as was now crystalizing in perceptive people's minds, came the realization that should the Union resort to "coercion" to stamp out the fast-spreading fires of

Southern rebellion, its armies would need to march through Virginia to do so. In that light, remaining in the Union or even attempting to adopt a stance of neutrality—if such a thing were possible—brought significant risks of war. To secede would ensure that the first battle would be fought in Virginia.

All sides understood that Virginia's addition to the fledgling Confederate republic would be hugely beneficial. Not only did the Old Dominion boast a population larger than any Southern state, it also possessed more slaves and had a greater acreage of land under cultivation than did any of its Confederate counterparts.[3] Its exposed position pressing up against the Northern states was, to fire-eating Southerners, a big advantage if hostilities erupted—with Washington a stone's throw across the Potomac, Virginia's secession would bring the federal vipers' nest within reach of its arms.

As Virginia's voters streamed to the polls on that historic 4th of February, one wonders how many knew just how high the stakes were. Whatever decision the Old Dominion made was fraught with peril.

* * *

Meanwhile, the 132 delegates to the Washington Peace Conference were beginning their deliberations. The brainchild of the former president John Tyler and endorsed by the legislature of his home state, Virginia, the conference's opening date of 4 February had been purposely chosen to overshadow the secessionists' gathering in Montgomery. But the national response was less than enthusiastic. Only twenty-one of the thirty-eight states were represented—no deep South or West Coast states had, or could have within the time available, sent delegates. And those who did show up were largely past-their-prime "old gentlemen" who wasted three valuable weeks coming up with a peace plan that, more or less, replicated a compromise first proposed by Kentucky's Senator John Crittenden. It was an uninspired performance, hardly calculated to avert the impending catastrophe of disunion.[4]

The Virginia state delegates, who first met in Richmond on 13 February, began the task of recommending a future course for the Old Dominion. It quickly became clear that unity of vision between the state's disparate geographic regions over the matter of secession did not exist; the voters had ensured that outcome. The voters had also stipulated by a 2:1 margin that whatever formula the convention adopted, it must be submitted to a state-wide referendum. Of the 152 elected representatives who jostled for space that day inside the Virginia Capitol building's House of Delegates chamber, a small minority were in favor of outright secession, another equally zealous minority pressed for even closer bonds with the Union, while the greatest majority desperately searched a formula that would preserve the Union and be broadly acceptable to most Virginians. One favored concept which was repeated with multiple variations was that of a border state conference, to include only those slave states still within the Union like Virginia. But the hard work of compromise soon gave way to grudging procrastination. As one historian acidly observed, "Endless speeches comfortably avoided uncomfortable choices." Virginia's experience mirrored that of the nation.[5]

While the delegates' numerous discourses may have been overly long, many contained priceless insights into the issues that fractured the nation. One such pro-secession oration was given by George Wythe Randolph, a delegate representing Richmond and the youngest grandson of Thomas Jefferson, who would later briefly serve as the Confederacy's secretary of war:

> … On the 6th of November last the Government passed, in strict accordance with Constitutional forms, into the hands of a sectional party, organized upon principles hostile to our institutions, social, political and industrial. In the opinions of some, the observance of Constitutional forms in that election robs it of all wrong. In my judgement, it is that very observance of form which renders the election so dangerous to our liberties. Had revolution been attempted, we … should have opposed with the strong arm of force the seizure of the government. But the fact that our enemies found it unnecessary to violate the forms of the Constitution proves the insufficiency of that instrument for our protection … The Government, then, if I may so express it, is constitutionally revolutionized, and requires a counter revolution to restore it.[6]

* * *

Virginia's current governor, John Letcher, like many of his peers, had first pursued a career as a lawyer. His bookish, bespectacled looks gave him more the appearance of a professor than a politician. Hailing from Lexington, Letcher was elected as a Democrat to serve his district in the U.S. House of Representatives and did so for six years. He then ran for governor and won, taking office on 1 January 1860. John Letcher's positions toward the existential issues facing Virginia were fluid; he was not an abolitionist but he also discouraged secession. As a moderate politician, he tended to move with the voice of the people.[7]

One could never make that same claim about Letcher's predecessor, Henry A. Wise. Virginia's Constitution forbade a sitting governor to again run for office, so Wise dutifully if reluctantly returned to his former lawyerly pursuits. But politics ran deep in his blood and so he stood for election as a Virginia State Convention delegate from Princess Anne County, just to the east of Norfolk. Together with his fellow delegates, Henry Wise took his place in Richmond's Capitol building on the 13th of February.[8] And of all the men who gathered together that day, none would exert a greater impact on the fate of Virginia and the crisis that enveloped the Gosport Navy Yard than former governor Henry Wise.

Of Virginia's prominent men, few claimed roots in its Eastern Shore, an isolated peninsula separated from the state's more familiar Tidewater region by the fluid divide of the Chesapeake Bay. It was there, in Accomac County, that Henry Wise was born in 1806. The fifth child of a family of modest means, young Henry was privately schooled and then sent to a local academy before leaving to attend college and law school. His upbringing in a settlement surrounded by water and accessed almost exclusively by boat made its mark on Henry Wise, for it taught him at an early age to handle under sail any of the local craft.[9]

First practicing law, Wise gravitated toward politics. At age twenty-six he was elected from his home Eastern Shore district to the U.S. House of Representatives. His early leanings were those of a strong Jacksonian Democrat but he later broke ranks and took a more independent stance aligned with the Whigs. Before many years passed, the youthful Henry Wise had won a reputation in Congress as one of the nation's leading orators. His magnetic personality and ability to deliver eloquent speeches led one newspaper editor to opine that "the popular heart will leap toward him like the flames of fire to their native heavens."[10]

During his years in Congress, Wise was appointed to the House Committee on Naval Affairs, eventually rising to become its chairman. Through his efforts, in 1842 a bill was passed that funded the construction of the Stevens battery, the Navy's first ironclad warship. A screw-propelled, iron-hulled steamer that was technologically ahead of her time, she was never completed and languished on the stocks for decades. But at the time of her inception, Henry Wise had the sagacity to recognize Stevens' creation as the future of naval warfare.[11]

In his appearance, Henry Wise was hardly the caricature of a Virginian gentleman. He chewed tobacco, was often vulgar and had an explosive temper. His sharp, gaunt features, penetrating eyes and long lank hair gave him an almost dangerous look. Thin, wiry and animated by nervous energy, his dignity rose from the sheer force of his personality and his integrity of spirit.[12] As his grandson and biographer, Barton Wise observed:

> If you judge Mr. Wise by the acts of his life ... he has few claims to consistency. But if you judge him by the impulses of his nature and the fidelity and chivalric bravery of his adherence to them, the verdict in his favor is emphatic and beyond question.[13]

Henry Wise married three times. His first wife was killed in a fire and his second wife died during childbirth. He sired fourteen children of whom seven survived their infancy.

After five terms in Congress, Henry Wise accepted an appointment by President Tyler as the U.S. minister to Brazil. In May of 1844, Wise and his family took passage on the frigate *Constitution* for Rio de Janeiro, his home for the next three years. They returned to America on the frigate *Columbia*, both voyages giving Wise, a skilled boat-handler, experience on board a living warship and a first-hand chance to observe the Navy's way of doing things.[14] Throughout his antebellum career, Wise never actively served in the U.S. military although he clearly displayed an abiding interest in all things naval.

Not content with the quiet life of a lawyer, Henry Wise actively participated in Virginia's convention of 1850-51 that led to significant updates to its Constitution. Wise's embrace of both Democrat and Whig political philosophies enabled his election as Virginia's governor in 1855. It was during Wise's tenure that John Brown launched his infamous raid on the federal arsenal at Harpers Ferry. To Governor Wise fell the decision to approve or commute Brown's death sentence. After interviewing Brown in his prison cell, Wise was convinced of his sanity and allowed the execution to proceed. In doing so, Governor Wise made a strong statement about the Old Dominion's stance on abolition. Commanding the state militia charged with maintaining order during Brown's trial and hanging was Wise's old friend and Mexican War veteran Major General William B. Taliaferro, a Virginian who will be heard from again.[15]

With his term as governor drawing to a close and years of retirement lying ahead, Henry Wise decided to move to a more accessible location than his long-held Accomac farm. In the fall of 1859, he purchased his elder brother's estate, Rolleston, a fine property of 884 acres located in Princess Anne County, just seven miles from Norfolk on the eastern branch of the Elizabeth River. This meant that whenever Wise traveled to or from his new home, he would unavoidably first pass through Norfolk and, whether journeying on by steamer or railroad, catch a glimpse of the bustling waters fronting the Gosport Navy Yard.[16]

* * *

It is hardly surprising that the increasing polarization of the North and the South captured Henry Wise's attention. Although now out of office, Wise delivered a speech in Norfolk during the fall 1860 campaign season that laid out his views of how Virginia should position herself in the event Abraham Lincoln was elected. After first questioning whether the Old Dominion would rise to the challenge, he stated:

> If she does not meet the issue and come up to the mark of self-defence and self-respect, I will look to another and another and another, until some one sovereign does raise the rightful flag of Revolution. Revolution is the word. ... Secession is revolution, but revolution is not secession. I will not nullify, I will not secede, but I will under sovereign State authority fight in the Union another revolutionary conflict for civil liberty, and a Union which will defend it. ... If it must come, let it come as soon as possible.[17]

Should any of the citizens in his audience have wondered whether he would soon be raising his sword, Wise offered these words of comfort: "I will wage no private war. I will take part in no unauthorized foray. I shall first await the action of my own sovereign state."[18] But if the revolution beckoned, would the impulsive former governor remain true to his pledge?

Henry Wise's novel doctrine of "fighting in the Union" was based on his interpretation of the Constitution's Article I, Section 10, Clause 3, which permits a state to keep troops and ships of war in times of peace and to engage in war when "actually invaded, or in such imminent Danger as will not admit of delay." To Wise's way of thinking, the Old Dominion could most effectively protect its vital interests and avoid a collateral slide into war by raising a mailed fist and asserting its rights under the cover of the Constitution. This meant neither seceding nor allowing the Union to trample its rights or march across its soil, but an independent middle course of armed neutrality. Obscured in the smoke of an accelerating wave of secession and the Union's disjointed reactions, Wise's doctrine failed to gain widespread recognition.[19]

During this same fall campaign season, while lecturing in his new home county Henry Wise proposed the organization of local "committees of safety," presumably modeled on those that sprang up during the Revolutionary War. His call struck the wrong nerve. Wise's opponents, fearful of vigilante violence, branded him as "treasonous" and an enemy of the "power of the people."[20] The strong reaction indicated that pro-Union sentiment still smoldered in a region that hosted a large navy yard and was thankful for the hundreds of jobs it provided.

Another more radical scheme aired by former Governor Wise was a harbinger of things to come. Voicing his concerns that the federal facilities within Virginia's borders could metastasize into "free soil enclaves," he raised the cry in December 1860 for capturing Fortress Monroe while it was still relatively defenseless and weakly garrisoned. Other places of interest included the arsenal at Harpers Ferry and the Gosport Navy Yard. It is said that Wise was part of an organization formed to seize Fortress Monroe but was forestalled in part by worries that if successful, Governor Letcher would not support such a radical action.[21]

It can be grasped that Henry Wise owned very definite and distinctive ideas of just how Virginia should react to the impending crisis of civil war. He was widely respected, dynamic and unstoppable. His call for "fighting in the Union," his urging of aggressive citizen actions against federal authority and his participation in Virginia's Convention to consider secession stood tall as mileposts on Henry Wise's personal road to revolution: "I say, then, stick to all your rights, renounce none, fight for all, and save all!"[22]

* * *

Meanwhile, in Washington, time was running out. With each passing day the date for the Lincoln administration's inauguration, 4 March 1861, drew nearer. After weeks of closed sessions, the old gentlemen of the Peace Conference finally wrapped up their labors—a seven-part amendment to the Constitution—and presented it to Congress. However, they did so on 27 February, just three days before the end of its lame-duck session. By this late date Senator Crittenden's very similar compromise, dealing with the extension of slavery into new territories and constitutional guarantees to preserve slavery where it existed, had been recast as a resolution for an advisory plebiscite. But the novelty of the legislation soon mired it in procedural complications.[23]

During these same quickly passing months, the composition of Congress had subtly shifted. The seats held by Senators and Representatives from seceded states stayed conspicuously empty, increasing the power of those states remaining within the Union. Amplifying this effect, the Congressional Republicans stuck together and refused to compromise, their gaze firmly set on the 4th of March when they would control the government. They were unwilling to abandon their party's primary plank—opposition to slavery—or make any concessions under pressure that would hold their new administration hostage. As the future secretary of the treasury, Salmon Chase saw it: "Inauguration first—adjustment afterwards."[24]

In the end, hopes for a legislative solution to the nation's secession crisis steadily diminished and then swiftly evaporated. The House adjourned on the 3rd and yet the Senate staggered on into the morning hours of inauguration day. Before it, too, adjourned votes were cast on both the Peace Conference plan—it failed by 28-7—and on the Crittenden compromise, which was defeated by the narrow margin of 20-19. During this legislative final gasp to forestall secession, the Republicans in Congress never gave their majority vote to any of the proposed compromise measures and even more starkly, not a single Republican vote was cast in favor of Senator Crittenden's compromise plan. The exhausted legislators had failed, but it would be more correct to say they never really had a chance of succeeding.[25]

Anxious citizens of the border states read the news from Washington with a sense of gloomy foreboding. Any hopes that an amicable national reconciliation was within reach were dashed. What course would the people of Virginia, North Carolina, Tennessee, Kentucky, Missouri or Arkansas take? Would they conclude that the nation was irrevocably fractured and throw in their lot with the Confederacy? All eyes and ears turned toward the coming day when the United States' sixteenth president, Abraham Lincoln, would be sworn in and address his fellow citizens.

Many also gazed toward Richmond, seeking clues in the deliberations of its State Convention. Would the Old Dominion stay in the Union or secede? The fate of the nation was in the balance.

* * *

CHAPTER 5

Lincoln Takes Command

Abraham Lincoln stood solemnly on the east portico's inaugural platform, his prepared speech in hand, a teaming crowd of black-clad citizens stretched out before him, eagerly anticipating his address. The Capitol's unfinished dome rose awkwardly behind him, its tall crane now stilled, suggesting to an observer that something was amiss within America.[1]

The nation's newest president first offered words of reassurance, pledging to not tamper with the institution of slavery or to send troops into the seceded states. He would respect all existing laws. His audience now reassured, Lincoln ramped up his tone. The Southern states' ordinances of secession were, he claimed, "legally void," they had no right to break away from the Union, and their attempt to do so was an act of insurrection. He branded secession as "the essence of anarchy." Further, Lincoln stated that his administration would "hold, occupy, and possess" federal forts and facilities in the seceded states. In concluding, he invoked the famous "better angels of our nature" phrase, rounding out an address that he hoped would be received as "firm but defensive and pacific."[2]

By claiming that no state could lawfully break away from the Union, Lincoln was, as were the seceded states, wrestling with an existential contradiction in the United States Constitution and its grant of federal authority overlaying a matrix of states' rights. In a letter to Virginia's Henry Wise written some years later, former mayor of New York City, Fernando Wood, diagrammed the flaw that had sundered the nation:

> The Republic form "is that of a State in which the sovereign power is in the representatives elected by the people." This was not nor is yet our system according to the theory of State government. The States are republican, but not the Federal Union. Two sovereign powers cannot exist within the same government. If sovereignty was

> retained by the States, it could not have been imparted to the Federal authority. Opposite views as to the relative strength of these authorities have existed ever since.[3]

Abraham Lincoln's inaugural address firmly declared that federal authority was sovereign throughout the United States. However, the loss of the central government's grip was substantial and growing. Seven states had already seceded and others had begun the process. The two Union bastions within the newly created Confederacy were in urgent need of relief. Fort Sumter had been under siege for nearly ten weeks and Fort Pickens for seven. Just how Lincoln's administration would execute the president's proclaimed policy and enforce his vision of federal supremacy was far from clear.

It was a delicate balance Lincoln needed to strike. If he harshly enforced his bedrock commitment to preserve the Union, it would drive the border states into the arms of the secessionists. If Lincoln trusted good faith and compromise, he risked entrenching the anarchy of secession he so feared. The seceded states would then be even more immune to pleas for national reconciliation. Abraham Lincoln's task would have been daunting for any government; for a newly installed administration, it verged on overwhelming.

* * *

To pursue his presidential mandate, Lincoln apportioned the responsibility for implementing his policies among a cabinet of five secretaries (state, treasury, interior, war and navy), his attorney general and the postmaster general. The one cabinet member whose actions most profoundly impacted the fate of the Gosport Navy Yard was the secretary of the Navy. Lincoln's pick for this position was Gideon Welles.

Welles' appointment had a lot to do with a president's need for sectional and political balance within his cabinet. Gideon Welles of Connecticut represented the minority Democratic element of the Republican Party and, because of his native state, could claim a New England constituency with its powerful maritime interests. He came to his secretary's office with more experience than any of his predecessors, having served as the Navy's chief of its Bureau of Provisions and Clothing for three years spanning the Mexican War. Although Welles never actively served with the United States military, he knew the ways of war-time Washington and the Navy.[4]

Fifty-eight years old when he took up his office, Gideon Welles had been educated as a lawyer. His early career was in newspapers—he was the editor and proprietor of the *Hartford Times and Weekly Advertiser*—and in politics. He was a member of the Connecticut state legislature and obtained executive experience while serving as Hartford's postmaster. Originally a Jacksonian Democrat, Welles converted to the Republican Party following the divisive passage of the Kansas-Nebraska Act and sat as a delegate to their 1856 and 1860 national conventions. It was at the latter that he threw his support behind the dark horse candidacy of Abraham Lincoln, winning gratitude that led to his secretaryship.[5] Welles was married and had eight children, only three of whom lived to adulthood.

As his extensive correspondence and multi-volume diaries illustrate, Gideon Welles was a gifted writer, clearly able to convey complex thoughts and clear logic. He was quiet, solid and not showy—the exception being the flowing wig and full whiskers he wore, which gave rise to his nicknames of "Father Neptune" and "The Old Man of the Sea." As an executive, he followed the sensible protocol of assembling the facts, asking probing questions and justly administering the necessary actions. Gideon Welles went on to serve eight years under Presidents Lincoln and Johnson and would go down in history as one of the best secretaries of the Navy. However, it can't be emphasized enough that although he was relatively well prepared for his office, like all of the incoming Lincoln administration, Secretary Welles was new to his job.[6]

* * *

The Navy Department that Welles headed was not large, totaling less than sixty employees. Welles, his multiple clerks and messengers presided over five bureaus, each managing the self-explanatory naval functions implicit in their names: Yards and Docks; Construction, Equipment and Repairs; Provisions and Clothing; Ordnance and Hydrography; and Medicine and Surgery. His staff was housed mostly in the old Navy Department building, just to the west of the Executive Mansion (White House) facing 17th Street, with some officed in the nearby Winder Building.[7]

The United States Navy was still on a peacetime footing, its size limited by Congressional appropriations which had failed to increase in response to the threat of secession. As of 1861, the *Navy Register* listed a total of ninety vessels of war in the Navy's inventory, of which only forty-nine were in commission. The total was inflated by six warships still on the stocks—one at Norfolk—and included just twenty-six active steamers. All its battleships were sailing vessels, most decades old, as were its ten large frigates and twenty-three sloops-of-war and brigs. The Navy's only ironclad was the Stevens battery, which commenced in 1842 but as yet unfinished.[8] At first glance, it was hardly the image of progressive naval development, especially if compared to its European rivals.

But there were bright spots. Five of its latest vessels-of-war were powerful, screw-propelled steam frigates, the *Merrimack* class, which could mount sixty guns and best any existing ship of their type. Many considered them a match for two-deck liners, such as the Royal Navy's *Agamemnon.*[9] Due to budget and manpower constraints, they had been laid up "in ordinary" but could be rapidly recommissioned. Five of the Navy's first class and seven of its second class steam sloops had only recently been placed in service and were armed, as were the big steam frigates, with advanced shell-firing guns designed by John Dahlgren, now a commander. Even its fleet of sailing frigates, sloops and brigs gave value to a Navy charged with protecting a far-flung "empire of commerce" sustained by merchantmen that relied largely on wind power.[10] The commodious hulls of the Navy's obsolete but still sound ships-of-the-line had recently been surveyed and proposals aired by Secretary Welles' predecessor for their conversion into steam frigates (see Appendix 6). Whatever its strengths or weaknesses, the navy Gideon Welles inherited had been shaped by the wars it fought, nearly all against foreign foes. It was a blue water cruising navy, ship-for-ship one of the best. But if President Lincoln so commanded, it was a navy ill-suited to exert control over the waters lapping its own shores.

During times of peace, the U.S. Navy's assets had traditionally been forward deployed to distant cruising stations in fulfillment of its mandate to protect American commerce. The Navy's first true foreign service squadron prowled the Mediterranean Sea in response to depredations by the Barbary pirates. Since then, the West Indies, Pacific, Brazil (or South Atlantic), East Indies and African stations had been added. The squadrons cruising these waters operated far from home and their ships—traditionally several sloops-of-war, a frigate or two, and in some cases, a liner—kept the seas for years on end before returning to a U.S. Navy yard. In an age of inefficient steam engines and given the necessity of patrolling remote coasts with few, widely spaced coal depots, canvas powered ships multiplied the effectiveness of the few steamers available to foreign stations. And nearly all those ocean-going warships, both sail and steam, had hulls with drafts deeper than most Southern ports could accommodate.[11]

The Navy's Home Squadron, which guarded waters adjacent to the U.S. eastern seaboard, was ironically the newest grouping of its fleet. Formed in 1841, it was a response to war scares with Britain and subsumed the former West Indies Squadron. When Gideon Welles came to office, a significant element of the Home Squadron was deployed off the port of Veracruz with some recently redirected to Pensacola. Mexican government instability together with concern that it might default on its foreign debt prompted the State Department to request an American show of force.[12]

The flag officer of the Home Squadron, Garrett Pendergrast, lately flew his broad pennant from the sloop-of-war *Cumberland*, one of several vessels deployed to Veracruz for that purpose.

* * *

Secretary Welles had a far more serious matter than the capabilities of the Navy's ships or their distribution among squadrons to worry about—the morale of his Department. As he observed in his diary, "A strange state of things existed at that time in Washington. The atmosphere was thick with treason."[13] Not only were the civilian personnel in his office of questionable fidelity, but even more alarmingly the specter of secession had prompted an increasing number of Southern officers to resign. Welles had vocally criticized his predecessor, Isaac Toucey, for so readily accepting their letters of resignation.[14] By the time Gideon Welles assumed his secretaryship, over seventy resignations had been accepted in contrast to the harsher response, dismissal from the service, which had been invoked just once. The outgoing House of Representatives had reached the same conclusion. Before adjourning they voted to censure Toucey for accepting so many resignations "without delay or inquiry."[15]

In response to his need to appoint loyal and capable officers to the never-ending rotation of naval assignments, Secretary Welles created a new function, the Office of Detail.[16] Headed by a solid and reliable officer, Captain Silas Stringham, the office relieved Welles of much of the burden of vetting officers and drafting their necessary orders. Compared to Toucey, Welles' attitude toward weeding out disloyal officers was one of stringent and zealous efficiency. Each serving naval officer was now required to reaffirm his oath of loyalty to the United States.[17]

A less visible effect of the infidelity lurking within the Navy Department was the lack of security it could exert over any of its documents or communications. The upshot was that "many of the most important orders given at that early day were verbal, unwritten instructions."[18] This precaution necessitated the dispatch of a reliable officer to repeat vital orders to another distant officer, yet one more burden on an officer corps already suffering from the loss of key personnel, all during an uptick in operational tempo. An even more pernicious consequence was to rob future scholars of the exact wording of many important orders issued at critical moments, the only accounts of them appearing later in testimony or memoirs.

During his first week in office, amidst an atmosphere of intrigue, treason and uncertainty over the Union's future, Secretary Welles re-examined the findings of Captain Armstrong's court of inquiry over the loss of the

Pensacola Navy Yard. His predecessor had chosen to do nothing but Gideon Welles was cut from a different cloth. On 12 March 1861, just seven days after he was sworn into his office, a general court-martial was convened in Washington to try Armstrong on the charges identified by his court of inquiry.[19] One can surmise that Welles' motivation rose from the same harsh judgment he applied to the officers' letters of resignation piling up on his desk. By acting so quickly and decisively, Welles broadcast a message that he would not tolerate lapses of loyalty or of propriety of conduct.

* * *

It must not be thought that the plight of the Gosport Navy Yard had escaped the attention of President Abraham Lincoln or Secretary Welles. However, their focus on Norfolk was necessarily filtered through the lens of the larger issue of Virginia and whether she would secede. Welles described the administration's reasoning:

> At the time of Mr. Lincoln's inauguration, and for several weeks thereafter, he and others indulged in the hope of a peaceful solution of the pending questions, and a desire, amounting almost to a belief, that Virginia and the other border States might, by forbearance and a calm and conciliatory policy, continue faithful to the Union.[20]

The specific measures adopted by the Lincoln administration to promote its non-coercive policy were entirely passive. They would later come back to haunt Secretary Welles:

> The President desired that there should be no step taken which could give offence; and to prevent any cause of irritation, he desired that not even the ordinary local political changes, which are usual on a change of administration, should be made. In regard to the navy-yard at Norfolk, he was particularly solicitous that there should be no action taken which could indicate a want of confidence in the authorities and people, or which would be likely to begat distrust. No ships were to be withdrawn, no fortifications erected.[21]

In his next breath, Welles stated that he had received "reports from that station ... that there were ardent secessionists among the civil and naval officers ..." but Lincoln's prohibition against making any changes of personnel, especially those driven by political allegiance, forbade taking remedial action. Recall Captain Barron's January report about the hundreds of workmen parading about the Gosport Navy Yard wearing the blue cockade of secession. But more ominous than disaffected civilian workers were the yard's unfaithful civil and naval officers; they were by far the greater menace. But Welles' hands were tied. Soon after taking office, Lincoln made a "special request that no important or extensive changes should be made in the navy yard at present without consulting him."[22] But Lincoln's edict didn't prohibit Secretary Welles from checking the bona fides of Charles Stewart McCauley, the yard's commandant:

> Commodore McCauley, who was in command of the Norfolk yard, I had personally known in former years, and esteemed as a worthy and estimable officer. His reputation as a Union man in 1861 was good, and all my inquiries in relation to him were satisfactorily answered. His patriotism and fidelity were beyond doubt.[23]

Whether the administration's policies had been spelled out to Commandant McCauley is a question for speculation. No such communication of policy can be found in the *Official Records* nor do they contain any indication, until much later in mid-April, of a visit to McCauley by an officer bringing verbal instructions. Secretary Toucey had dispatched a personal messenger to Norfolk in the shape of Captain Levin Powell, but apparently, Gideon Welles felt no need to do so during his first weeks in office.

An opportunity for Welles to take McCauley into his confidence arose on 11 March, the day before Armstrong's court-martial began. Commodore McCauley had sent a report to the secretary advising him that "the sloop of war *Germantown* will be ready for her officers and crew by the 25th instant."[24] Such reports from a commandant were routine and gave the Navy Department advance notice of the necessity to gather and dispatch the requested personnel. In the case of *Germantown*, nearly 200 crewmen were required, all of whom would need to be recruited (if not already enlisted) and sent on from a northern seaport, typically New York. Sailors in these numbers were simply not available in Tidewater Virginia. Her compliment of some fifteen officers also had to be detailed for duty.

There is no evidence that Secretary Welles issued the necessary orders to address Commodore McCauley's advisory notice or that he acknowledged it.[25] Perhaps Welles held back from ordering a crew for *Germantown* to avoid clashing with Lincoln's policies. It would mark the first step in removing her from the navy yard. Or he may have been denied permission to act on a request that might have been construed as fortifying Norfolk. A note of irony clings to Welles' lapse. Just six weeks earlier, McCauley proposed to Captain Powell that the crews for two ships preparing for sea should be sent to him as an ad hoc reinforcement of Gosport's defenses. And now, when McCauley's need for a crew was genuine, his communication appears to have fallen on deaf ears. At a time of growing concern over the security of the Navy's most valuable dockyard, its commandant's sensible request should have been a top priority.

* * *

President Lincoln may have devised a coherent if flawed plan to cope with uncertainties over Virginia, but he found it far more difficult to achieve a unity of opinion among his cabinet over the larger issues driving the nation's crisis. William Seward, Lincoln's secretary of state, figuratively stood apart from his cabinet colleagues, inflated by his belief that he was "the one indispensable man" during the country's hour of need. Leveraging warm personal ties formed during a long political career, throughout the turmoil of secession Seward privately maintained links with many of the Southern leaders. Now he espoused a program of "forbearance, conciliation, magnanimity" toward the Confederacy, preaching that any attempt to exert the Union's sovereignty by force would backfire and cause the border states to "go South." In practical terms this meant sacrificing Fort Sumter, which many in Lincoln's cabinet were coming to accept, but balancing that loss—and so saving face—by holding on to the less confrontational of the threatened federal enclaves, Fort Pickens.[26]

Secretary Welles did not see eye to eye with Seward. One an experienced political operative from Connecticut, the other a wily former Empire State governor and senator, they would have naturally clashed over lingering party animosities. But the matter of secession was charged with a higher passion. Welles, who was regarded by many Southern sympathizers as "one of the war wing of the Republican party," strongly believed Seward's back channel machinations were "inappropriate if not treasonous:"[27]

> Seward ... was, at that moment, ... carrying on an intrigue with the Rebel leaders, who were deceiving him, whilst he flattered himself that he was using and could control them."[28]

The one member of Lincoln's cabinet who consistently and resolutely resisted the drift toward accepting Fort Sumter's surrender was Postmaster General Montgomery Blair. By mid-March, even the U.S. Army's General Winfield Scott had recommended the evacuation of Fort Sumter and was wavering on whether Fort Pickens could be held. Blair, a West Point graduate, would have none of it and "took the opposite view from Mr. Seward and General Scott .... his indignation that any idea of abandoning

Sumter should be entertained or thought of was unbounded."[29] An invisible operative hovering in Blair's shadow was his brother-in-law, an energetic retired naval officer with the bigger-than-life name of Gustavus Vasa Fox. Former Lieutenant Fox had the month before proposed to the Buchanan administration a carefully calculated scheme for fast tugs to dart into Charleston's harbor and perform what military experts had deemed impossible, relieve Major Anderson's garrison.[30]

But time was running out. As all the nation knew, Anderson's supplies would be exhausted by mid-April. A decision not to relieve him would be inexorable, the quick march of time refusing to permit reconsideration. On the other hand, a decision to attempt relief could be temporized at any time, allowing hope to still burn. The tipping point for President Lincoln was the very real threat of Montgomery Blair's resignation:

> Before handing in his resignation, a delay was made at the request of his father. The elder Blair sought an interview with the President, to whom he entered his protest against non-action, which he denounced as the offspring of intrigue. His earnestness and indignation aroused and electrified the President; and when in his zeal, Blair warned the President that the abandonment of Sumter would be justly considered by the people, by the world, by history, as treason to the country, he touched a chord that responded to his invocation. The President decided from that moment that an attempt should be made to convey supplies to Major Anderson and that he would reinforce Sumter.[31]

Following a pivotal cabinet meeting held on 29 March, during which all except Seward agreed to attempt to save Sumter, President Lincoln issued orders to prepare a relief expedition. Gustavus Fox's daring plan was on. But Seward refused to let go of his goal of reinforcing Fort Pickens and pressed for a scheme centered on the Army's Captain Montgomery Meigs. Two days later, Lincoln agreed. At first blush, all looked promising. Lincoln had commanded that Forts Sumter and Pickens be saved. Two separate relief expeditions were now approved at the highest level. But while Fox's expedition would be organized through regular War and Navy Department channels, the force led by Meigs fell under a shadowy command structure devised by the secretary of state.[32] Their success depended on careful joint planning. But as with any military action, the devil was in the details. And those messy details would lead to a world of unintended consequences for the Gosport Navy Yard.

* * *

CHAPTER 6

Cumberland Arrives at Hampton Roads

It was Saturday the 23rd of March and all eyes were on the tall sloop-of-war making for the anchorage. Under topsails and topgallants, her crew swarming her deck ready to moor, the pilot detached himself from a cluster of officers and peered over the weather rail one last time. *Cumberland*'s bow was aimed straight at the berth he pointed to. Her way now checked, she made a running moor, dropping her bower anchor and paying out chain while the well-braced yards and close-hauled spanker helped her to round into the wind. All sails were quickly furled, lines neatly coiled, and the big sloop lay snug to her anchor as the sea breeze died away and the sun's bright orb rose high over the water. So arrived *Cumberland*, the U.S. Navy's Home Squadron flagship, twenty-seven days from Vera Cruz, flying the broad pennant of Flag Officer Garrett J. Pendergrast.[1]

Cumberland had been ordered to Hampton Roads by Secretary Toucey, and knowing that he would be out of office before she arrived, he added the sensible caveat to his orders for *Cumberland* to "there await the further instructions of the Department."[2] She was a magnificent ship, every inch a man-of-war. Built as one of the last "Gradual Increase" frigates, five years earlier *Cumberland* had been razeed: her upper works were cut down and the broadside guns removed from what was now called her spar deck. Her gun deck port spacing was increased to allow room for a larger, more powerful cannon. She now carried twenty-two of the latest Dahlgren 9" smoothbore shell-firing guns and mounted a 10" pivot gun on her forecastle. Manned by a complement of 350 officers, sailors and marines under the immediate command of Captain John Marston, she was one of the Navy's most powerful war vessels. Her one weakness, especially for a squadron flagship, was that she was propelled by sails, not steam. But bravely handled, she could give any foe a sound thrashing.[3]

The Home Squadron's flag officer, Garrett Pendergrast, was not happy flying his flag on a sailing ship and would soon request a steamer be appointed in her place. But the fifty-eight-year-old commodore was easily displeased by many things, which meant the ships he sailed in were prone to problems with discipline. Five years before, Pendergrast was appointed captain of the new steam frigate *Merrimack*—the very same vessel now under repair at the Gosport Navy Yard and the future object of so much angst—with orders to show her off in European ports during a special service cruise. Early on she suffered a propeller bearing failure and was forced to return to Boston for repairs.[4] There 169 of her crew—*thirty-five percent*—deserted.[5] When asked by the secretary for an explanation of his near mutiny, Pendergrast replied "we need look no further than the abolishment of flogging for a solution to the whole matter."[6] At *Merrimack*'s next port of call, three passed midshipmen requested to be detached. And his sailors kept deserting. An unhappy ship, indeed.

Happiness aside, Garrett Pendergrast had much the same profile as his fellow captains of the U.S. Navy. He cut his teeth as a midshipman during the War of 1812 and worked his way up through the ranks of the peacetime navy. His personal background, however, makes an interesting study, especially considering he was an active participant in the catastrophe that was to befall the Gosport Navy Yard. Pendergrast was born in, appointed from and a citizen of Kentucky, a slave-holding border state. At age twenty-four he married Susan Virginia Barron of Norfolk, Virginia, the daughter of Commodore James Barron of *Chesapeake-Leopard* infamy who later dueled with and killed Navy icon Stephen Decatur.[7] Susan's Barron ancestry also made her the niece of Captain Samuel Barron, the officer who wrote to Du Pont about Gosport's disloyal workmen. This Barron, a courtly Virginian labeled the "naval diplomat," was suspected of secessionist leanings; a month later he would "go South."[8] Susan's father the commodore died in 1851 and was buried in Portsmouth's Trinity Episcopal Churchyard, not far from the Gosport Navy Yard. She long maintained a residence in nearby Norfolk.[9] The Pendergrasts were childless but they adopted a son, Austin, who was now a lieutenant in the U.S. Navy. While nothing in Flag Officer Pendergrast's overt behavior gave cause to doubt his loyalty, his in-laws were awash in a strong Southern undertow.

* * *

Two days after Pendergrast's arrival at Hampton Roads, he penned a brief note to Secretary Welles:

> The *Cumberland*, having experienced a succession of severe gales off the coast, will need some repairs, and requires water, stores and

> provisions. I would therefore be glad to receive the orders of the Department in relation to them.[10]

This was quite a reasonable and proper step for him to take. In the meantime, several of *Cumberland*'s officers would have gone ashore to nearby Norfolk to arrange for the necessary stores and repairs. A visit to the Gosport Navy Yard no doubt occurred, and greetings were exchanged with Commodore McCauley and his officers. On 26 March, the day after Pendergrast informed Welles of *Cumberland*'s needs, McCauley also wrote to Welles:

> I would respectfully inform you that I have directed the U.S. ship *Cumberland* to be brought up from Hampton Roads to the naval anchorage on this station, provided it did not conflict with any order from the Department. I deemed this expedient to save time and expense in supplying the deficiencies that exist on her.[11]

And here begins a "tempest in a teapot" which itself was a trifle, but the interactions between the two senior Norfolk area naval officers are most revealing. As background, McCauley was the ranking officer, seven on the captain's list compared with Pendergrast who was thirty-one. McCauley was commandant of the navy yard, Pendergrast flag officer of the Home Squadron.[12] Knowing of *Cumberland*'s wants, in a communication to Pendergrast also sent on the 26th, McCauley "directed" that she be moved closer to Gosport. The naval anchorage he mentioned was at the widest point of the Elizabeth River, off Hospital Point between Portsmouth and Norfolk. Here a ship could comfortably lie within easy reach of the navy yard and all its satellite facilities. The commodore was simply trying to speed things up. But Pendergrast dug in and sent a stiff refusal to McCauley. He quoted his original orders, dwelling on the caveat that he was constrained to await "further instructions from the Department," but added the confrontational phrase "or my superior officer." He then concluded that "under the circumstances, I do not feel authorized to move this ship."[13]

Three days later, on 29 March, Welles replied to Pendergrast's original query and instructed him to "proceed with the *Cumberland* to the naval anchorage, Norfolk." Two days later Pendergrast confirmed "in obedience to that order I have this day moved the *Cumberland* ..."[14] But six days had been wasted in a pointless exchange provoked by Flag Officer Pendergrast's refusal to admit that Commodore McCauley was his senior officer. Such a stand-off hardly augured the formation of a close and confiding relationship between these two officers, one that would be vital to the future safety of the Gosport Navy Yard. It wasn't until eight months later that Pendergrast's ultimate dissimulation surfaced when he testified under oath to the Senate committee that later investigated the disaster that had befallen McCauley's navy yard.[15] In answer to a question about a later movement of *Cumberland* urged by McCauley, he unhesitatingly replied "Being myself subordinate to Commodore McCauley, it was my duty to act under his instructions."[16]

* * *

In itself, a navy yard is no more than a collection of inanimate buildings, structures, docks and machines. To perform its intended function—building and repairing warships—the executive touch of its assigned officers is the vital ingredient needed to bring a yard to life. Without them, it is but a rudderless shell. Each officer filled a specific role and in combination made a whole cloth out of its many threads. As the final days of March 1861 slipped by, the officers of the Gosport Navy Yard focused their individual skills on remedying *Cumberland*'s deficiencies.

First among the yard's line officers reporting to Commodore McCauley was Commander Robert G. Robb. A Virginian, Robb had been the yard's executive officer for nearly three years. Before his assignment to Norfolk, Robb commanded the sailing sloop *Cyane*. Under Commander Robb were three lieutenants, two from Virginia, Charles F.M. Spotswood and William Sharp. The third, Alexander A. Semmes, was a citizen of Maryland and a cousin of Commander Raphael Semmes, one of those Southern officers whose resignation Secretary Toucey accepted "without delay or inquiry."[17]

Gosport's substantial ordnance department was headed by another Virginian, Commander John R. Tucker. He had been at his post only ten months, but during the late 1850s Tucker commanded *Pennsylvania*, the yard's receiving ship. Many years before Tucker married Norfolk native Virginia Webb, the daughter of a Navy captain, casting him in a similar mold as Garrett Pendergrast. Tucker's wife had since died, but together with his many children, he kept a home on Norfolk's Fenchurch Street. Commander Tucker is a prime example of an officer who shaped his career around bouts of service at Norfolk to be close to his family.[18]

Serving under Tucker was the archetypical Virginian naval officer, Lieutenant George T. "Terry" Sinclair. The son of Commodore Arthur Sinclair of the War of 1812 renown, he was the younger brother of another Arthur Sinclair, a commander who was then captain of the yard's receiving ship.[19] Terry Sinclair had served as Gosport's gunnery lieutenant for nearly two and a half years. It was not his first time in this role; he had also filled it in 1856. Between stints at Gosport, Sinclair spent two years on the steam frigate *Wabash.*

Pennsylvania's captain, Commander Arthur Sinclair, had only recently arrived at Gosport. In November 1860 he was given the job of sailing the steam sloop *Pensacola,* newly completed at the navy yard of the same name, around to Washington for the installation of her engines, thereby keeping her from the secessionists. Sinclair's executive officer was the Norfolk-born Lieutenant Carter Poindexter. He was married with three children. The receiving ship's other lieutenants were Edward Donaldson, a Pennsylvanian who had reported to Norfolk six months before, and John Watters, a citizen of Michigan and newly arrived in January. Donaldson, like the yard's Lieutenant William Sharp, had previously sailed on the steam frigate *Merrimack.*

Responsible for Gosport's fleet of warships in ordinary—laid up and out of commission—was another Virginian, Commander Thomas Rootes. He had held his post for two and a half years, last serving at sea five years before on the sloop-of-war *Vandalia.*

The navy yard's rendezvous—its recruiting office—was in the hands of Commander Richard L. Page, a Virginia native and cousin of an up-and-coming Army officer named Robert E. Lee. A Gosport regular like Tucker, Page had previously served on the yard's receiving ship, as its executive officer and twice as an ordnance lieutenant. His assignment to the rendezvous had begun just five months before. He was married to Alexina Taylor, a Norfolk native, and had fathered four children.[20] Under Page was Lieutenant Peter U. Murphey, a citizen of North Carolina and the son of one of its state senators. He had been at his post for a year and a half and previously served in the yard as an ordinary officer.

Lieutenant Colonel James Edelin, a Maryland native, commanded Gosport's marine guard. He was married with two children. Reporting to Edelin was Captain Jabez Rich of Maine, stationed on the receiving ship, and First Lieutenant Adam N. Baker of New Hampshire, supervising in the yard. Under the two lieutenants, eighty-three marines stood ready for duty.[21]

A significant number of staff and warrant officers were also assigned to the navy yard and its facilities. Of the two surgeons and two passed assistant surgeons serving the Portsmouth Naval Hospital and in *Pennsylvania,* two were Virginians. Of the yard and receiving ship's six warrant officers, only the two sailmakers were *not* from Virginia. In contrast, the paymasters for both yard and receiving ship hailed from Northern states; only the yard's inspector of provisions was a Southerner.

Of Gosport's civil officers, navy agent George Loyall had resigned by this time. Naval constructor Samuel Pook had also been detached for special duty and was slated to be replaced by Portsmouth native John Luke Porter.[22] And attached to the yard on special duty was Chief Engineer Robert Danby, a citizen of Delaware.

A graphic summary of the Gosport Navy Yard's officer profiles is presented in Appendix 2, with updates for changes occurring closer to the cataclysm that struck in late April. Again, what jumps out is the overwhelming proportion of officers who were citizens of Southern states, most from Virginia, many of whom had settled down in the Norfolk area, were married to local women and had families living nearby.

But the Old Dominion's grip on the yard's senior commands was far greater than the percentages suggest—it was absolute. Gosport's executive officer, chiefs of ordnance, ordinary, rendezvous and receiving ship captain were *all* Virginians. After the Navy's experience losing Pensacola, and given the origins of Gosport's officers and the near certainty they would follow their states out of the Union, it strains credibility that Secretary Welles still trusted to "the President's policy of permitting things to continue undisturbed in Virginia."[23] Gosport was a secessionist powder keg waiting to explode.

* * *

In the wake of Lincoln's 29 March decision to proceed with two expeditions to relieve the North's beleaguered forts, orders flew fast and thick, first from the White House and then the Navy Department.[24] Their thrust propelled the rapid organization of a force afloat to support Fox's reinforcement of Fort Sumter. By doing so and thereby mobilizing the last of the Home Squadron's steamers forced Secretary Welles' hand on a matter that had been pressing him—Lincoln and Seward's urgings for its flagship, *Cumberland*, to return to Vera Cruz. Deploying the steam sloop *Pocahontas*, then readying for sea at the Gosport Navy Yard, for the Fort Sumter expedition coupled with worries about the adequacy of federal forces to "check disorderly proceedings, should any appear" in Norfolk, made it easy for Welles to argue that it was imperative for *Cumberland*, a sailing vessel, to stay put and act as a guard ship. Hence his 29 March order to Pendergrast to move her to the naval anchorage close by Gosport, as proposed by McCauley.[25] The two squabbling commodores probably had no inkling of the higher stakes driving the secretary's decision.

On the following day, 30 March, Welles sent messages to each of the navy yards where its steamers *Pawnee*, *Pocahontas* and *Harriet Lane* (recently transferred from the Revenue Service) were tied up, ordering them to be ready for sea "on or before 6th of April next, with one month's stores."[26] All three were earmarked for Fox's Fort Sumter expedition.

Now it was Gosport's turn for attention. The next day, 31 March, the secretary finally reacted to McCauley's long ignored request for *Germantown*'s crew and ordered Commodore Breese of the New York Navy Yard to "have in readiness for sea service without delay" three hundred "seamen, ordinary seamen and landsmen," to be transferred to Norfolk by a chartered steamer and on *Harriet Lane*.[27] Welles' entry in his diary explained his thinking:

> Notwithstanding the sensitive feeling that existed on the part of the people of Virginia, as well as of the Government, I felt that we might with propriety order a sufficient force there to man at least two of the smaller vessels without creating alarm, as it would be legitimate in the ordinary course of things. The *Plymouth* was designated as the practice ship for the midshipmen, and the *Germantown* was nearly ready for her armament and crew. No exception could be taken to orders to man them. If the seamen reached Norfolk, and an exigency should arise rendering it expedient to move the *Merrimac*, they could be made available for that purpose. The *Powhatan* had just reached New York and was ordered out of commission, but those of her crew whose time had not expired could be made available for valuable service at Norfolk.[28]

Taken at face value, the secretary's statement follows the logic and chronology of events, but only so far. But users beware: his diary's opening page contains a note alerting the reader to the fact that "This first chapter is not a part of Mr. Welles' diary, having been written several years after the events narrated ..."[29] So first, the agreement. Welles' plan to reinforce Gosport Navy Yard read exactly like Commodore McCauley's late January proposal to Captain Powell. Long since then, McCauley had notified the Department that *Germantown* would be ready for her officers and men on 25 March, a full week *before* Welles belatedly requested the transfer of a crew for her from distant New York. Now for his retrospective embellishments. The suggestion that *Plymouth* should be sent to the Academy didn't appear in the *Official Records* until 5 April, nearly a week *after* Welles ordered her crew, and the Department wouldn't order her reactivation until 11 April.[30] Similarly with *Merrimack*; her name only jumped into the correspondence on 10 April.[31] Perhaps the kindest reconstruction of Welles' rambling diary entry is that these future needs were tugging at the secretary, perhaps from verbal exchanges with officers at the Department, before they were formally put to paper. Whatever the reason, Commodore McCauley must have been comforted to hear his requested sailors were on their way to Norfolk.

Or were they? In an extraordinary volte-face, on 1 April Secretary Welles revoked instructions he had issued just four days before putting *Powhatan* out of commission, detaching her officers and transferring her crew, and now ordered that she be held "in readiness for sea service."[32] The big steam frigate had only arrived at the New York Navy Yard on 13 March after punishing service off China and then in the Gulf of Mexico and was in dire need of repairs and a refit.[33] Once her decommissioning order arrived on the 28th, it kicked off a well-rehearsed sequence of events for laying up a warship. Her powder was removed before she entered the yard, then once alongside all her stores were landed and, depending on her repairs, water tanks and coal bunkers emptied. Meanwhile, *Powhatan*'s officers removed their personal belongings, including cabin furniture, and all the Navy's loose items, such as her log books, nautical charts, almanacs, instruments, compasses, small arms and so on, were taken ashore for safekeeping. *Powhatan*'s flag was hauled down and she was turned over to the yard. Then her officers scattered, typically on three months' leave of absence, and her crew was discharged except those whose enlistments still had time to run; they were transferred to the yard's receiving ship. This was the state of things when Welles' order struck. And it was no April Fool's joke.

What could have motivated the abrupt change? *Powhatan* was later tagged to lead Fox's Fort Sumter expedition but was not one of the originally selected trio, *Pawnee*, *Pocahontas* and *Harriet Lane*. The decision for her inclusion wasn't made by Welles until after Lincoln gave Fox the final go-ahead on the afternoon of 4 April.[34] Orders to ready *Powhatan* for the sea with "all dispatch" and instructions issued to her captain, Samuel Mercer, outlining his cooperation with Fox, were only written on 5 April.[35] Examining the chronology of events and orders allows a speculative answer as to why Welles reversed *Powhatan*'s decommissioning on the 1st of April. It must have sunk into him, perhaps through urgings by concerned naval officers, that after earmarking *Pawnee*, *Pocahontas* and *Harriet Lane* for Fox's force, the Home Squadron's cupboard was empty. Thus, Welles' motivation was a derivative of Lincoln's original 29 March push for action: *Powhatan* became the Navy's final reserve. And, if ready in time, she could join the flotilla earmarked to relieve Fort Sumter. Welles later explained his dilemma:

> ... the Sumter expedition took from the Navy Department every available naval vessel. It was at this culminating period that vessels were most wanted in the Chesapeake and on the Potomac; for, in

> the case of conflict at Charleston, it was uncertain what would be the attitude of Virginia.[36]

But actions have unintended consequences and Secretary Welles' 1 April orders would reverberate far throughout the Navy's establishment. First to recognize the impact was the commandant of the New York Navy Yard, who the following day advised Welles:

> In consequence of the order to put the *Powhatan* out of commission being countermanded, I am unable to execute the previous order of the Department to send the draft of 300 men to Norfolk … as we now have no more men than will be required for the *Powhatan*.[37]

The New York Navy Yard, tucked inside the nation's most populous city, was overwhelmed by the sudden demand to recruit sailors for Norfolk, then *Powhatan* and *Harriet Lane*. Competing with the navy yard were the chartered merchant vessels outfitting for Fox's expedition, the enormous mail steamer *Baltic* and three tug boats. And other ships were preparing to sail for Fort Pickens. Men for all these were wanted in addition to those absorbed by the daily turnover of seamen in the ever-busy port of New York. Something had to give and it was the drafts consigned to Norfolk. Secretary Welles dutifully described the fallout:

> … But important events for the relief of Fort Sumter rendered it necessary to detain the seamen on the *Powhatan* for the Sumter expedition, and to add to them the recruits from the receiving-ship. These orders took almost all the recruits who were intended for Norfolk, as soon as two hundred and fifty were enlisted. Orders were given to … charter a vessel to take the men to Norfolk … but … could not be fulfilled.[38]

The sequence of Welles' orders during the four days immediately following Lincoln's go-ahead for the expeditions to Forts Sumter and Pickens exhibited confusion of leadership and a lack of foresight. Welles' biographer admits he "bungled" Gosport's relief operation and the secretary's missteps at the end of March were the first tangible evidence of this.[39] His abrupt reversal of *Powhatan's* decommissioning in the immediate wake of ordering the Navy's last three steamers to sea would have overstrained its resources. But inserting a concurrent transfer of 300 sailors to Norfolk into massive fleet mobilizations for the relief of two forts—overwhelming demands for a Navy still on a peacetime footing—

was the proverbial straw that broke the camel's back. The system buckled and the loser was the Gosport Navy Yard.

Welles' delayed decision to reinforce Norfolk was clearly driven by politics—the desire to appease Virginia—rather than by the good of the service. But think of what *might* have happened if the 300 men had been sent when they were requested. It would have made all the difference. Instead, Commodore McCauley was denied the crews he needed to activate the ships in his charge. Without men, they were inanimate hulks, hostage to fortune. *Germantown* was already on the block, with the sloop *Plymouth* soon to follow, and then the biggest prize of all, *Merrimack*.

* * *

CHAPTER 7

Virginia in the Balance

As March drew to a close, Virginia's State Convention floor still hosted its roster of impassioned speakers, be they Unionists, those who favored secession, or the rare individual like Henry Wise who pursued his eccentric vision of "armed neutrality." The Unionists held a fragile majority but their confidence in the schemes they proposed to resolve the festering sectional crisis was paper thin. A border state conference attended by all the slave states still within the Union made for a good speech, but could it achieve anything? Others who thundered that the North *must* agree to fundamental changes, such as those embodied in the failed Crittenden compromise, understood deep in their hearts that such ultimatums would likely get no further than the Mason-Dixon line. A reluctance to submit any of these grand schemes to the yes-or-no test of a vote had worked its spell among the delegates, encouraging mind-numbing procrastination.[1]

That Virginia's Convention had slowed to a crawl was readily apparent to anyone who could read a newspaper. The South Carolinian "fire eaters" had voted unanimously for secession three days after their convention had been convened. But Virginia's delegates hadn't reached any conclusions and had not voted on a single measure by the 1st of April, forty-six days after they had first met.[2] What were they waiting for?

Yet a hint of movement stirred the air. Seeking an understanding of where the meandering Virginia Convention was headed, the president and his secretary of state invited one of its prominent Unionist delegates to Washington. They had hoped that the Convention would quickly reject secession and adjourn, but that had not happened. So they sought an explanation. John Baldwin, a relatively unknown country lawyer from the Shenandoah Valley, was the chosen delegate escorted into the White House on the morning of 4 April. There he first met with Lincoln privately, where he stressed the themes that had underpinned the Virginia Unionists' position: removal of federal forces from the forts, sponsoring a national convention and support for guarantees of Southern rights. Rumors suggest that in reply, Lincoln may have offered to evacuate Fort Sumter in exchange for Virginia staying in the Union.[3] But the mood darkened when Baldwin picked up on the administration's plans to reinforce the fort. Baldwin later recounted "I told him that a relief expedition would cause guns to be fired." Regardless of who fired first, Virginia, "strong as the Union majority in the Convention is now, will be out in forty-eight hours." The president shook his head, saying that could not be so.[4]

On the same day Baldwin met with Lincoln, propelled by the urgings of an outspoken disunionist, Virginia's Convention steeled itself to vote on a motion tabled by Henry Wise's Committee on Federal Relations. It was an ordnance of secession and it failed by the substantial margin of 88 to 45.[5] Unionists should have been emboldened but the following day, after hearing the details of Baldwin's conversation with Lincoln, were deflated by the incipient march to war. Their tenuous position was revealed during acrimonious exchanges following a motion that three commissioners be appointed with instructions to visit the president and determine his true intentions. The only things the delegates could agree on were that any federal coercion by force would swiftly propel Virginia out of the Union and that they needed a cooling off period, and so the Convention adjourned until 8 April.[6]

Their spirits were revived by a quiet weekend, and the delegates soon got down to business. Now meeting again in Virginia's State Capitol after their temporary sojourn in the nearby Hall of the Mechanics' Institute, they agreed on the necessity of dispatching the proposed commissioners and nominated three men who represented the Convention's political spectrums: Alexander H. H. Stuart, an open-ended pro-Unionist; William B. Preston, a conditional or moderate Unionist; and George Wythe Randolph, an ardent secessionist.[7] But the commissioners were unable to start for Washington, their journey delayed by unseasonably heavy rainfall.

Meanwhile, in Richmond, the delegates' sense of foreboding increased as the rain thundered down and stories of Union expeditions filtered South. It must not be thought that the Convention's proceedings were a secret. On the contrary, they shared the newspapers' front pages together with headlines such as "The War Excitement" and "The Latest From Charleston." *The New York Times* published a report from Richmond dated 10 April under the title "The Virginia State Convention," in which it detailed, point by point, edits proposed to resolutions drafted by the Convention's committees. On that day, Henry Wise "moved to add a clause declaring that all the forts in the limits of the seceded states ought to be evacuated for purposes of pacification." His measure lost by 5 to 67. But the resolution, the twelfth passed by the Convention, although retaining its conditional language left no doubt where Virginia stood:

> The people of Virginia ... will expect, as an indispensable condition, a pacific policy will be adopted toward the seceded states, and that no attempt will be made to subjugate them to Federal authority; nor to reinforce any of the forts situated on the mainland, or within the harbors of any of the seceded states, nor to recapture the forts, arsenals or other property of the United States within their limits, nor exact payment of imposts on their commerce, nor any measures calculated to provoke hostilities or collision. The forts, arsenals, magazines, and other places ceded to the U.S. and situated on the mainland or for the protection of the harbors of the seceded states ought, for the purposes of pacification, be evacuated by the authorities of the Federal government.[8]

* * *

Back at the old Navy Department building in Washington, Secretary Welles struggled to keep ahead of the rapidly escalating events, all of which pointed toward war. As of early April, he had been in office for a month and now had a good grip on the routines of the Navy's business. In the course of his day's work, important matters were aired with visiting officers and the Navy's priorities were reviewed. What was the status of its squadrons? What work was in hand at the dockyards? And given the uncertainty over Virginia, what assets were at risk in Norfolk? Much of what occurred during such discussions likely went unrecorded; the documentation that exists yields at best an incomplete picture. But a glimpse into Welles' exchanges with one of his staff who voiced strong misgivings about Gosport can be found in a book published thirty-five years after the event, *The Steam Navy of the United States.* Written by Frank M. Bennett, a former passed assistant engineer and "favored with a thorough and most carefully written account of the affair from the chief actor," its author weaves a captivating story.[9]

On 25 March 1861 the Navy's engineer-in-chief, Samuel Archbold, unexpectedly resigned, motivated by a promising business opportunity. His replacement was urgently needed and probably at the recommendation of bureau chief John Lenthall, the following day thirty-eight-year-old Benjamin Franklin Isherwood, a chief engineer hailing from New York City, was appointed to fill his position, the highest an engineer could achieve in the U.S Navy.[10] Isherwood, carrying a reputation of being brusque, energetic and brilliant, quickly got to grips with the state of his fellow engineers' work. One of them, Robert Danby on special duty at the Gosport Navy Yard, submitted a report that captured his attention. Danby's biggest project by far was the steam frigate *Merrimack.* The year before her machinery had been condemned and was now undergoing extensive refurbishment, its parts scattered among the yard's workshops. Progress was slow. Even more disquieting, Danby, a loyal Union man, told Isherwood that the secessionists "counted surely on having *Merrimack* as the nucleus for their future navy."[11] Bennett recounted what happened next:

> One of the first acts of the new engineer-in-chief was to call the attention of the Secretary of the Navy to the possibility of getting the *Merrimack* away from Norfolk, and it is certain that had it not been for him no effort to that end would have been made. The Secretary's mind was engrossed with too many other important matters to give any thought to this particular subject unless it had been urged upon him and, it is more than probable … that he did not even know that *Merrimack* was at Norfolk.[12]

Also motivating Isherwood's "repeated urgings" of the secretary to act on *Merrimack*'s plight was the new engineer-in-chief's eagerness to become involved in issues at every level of naval administration. For most of his career, Isherwood, an engineer and therefore a staff officer, had struggled to achieve equality with the Navy's line officers—its lieutenants, commanders and captains. Engineering officers were not allowed to command ships or to represent themselves at their assimilated rank, only as naval engineers. Theirs was an inferior position, both professionally and socially, and Benjamin Isherwood was determined to do whatever he could to improve the lot of the Engineer Corps.[13] Typically staff officers (and they also included naval constructors, paymasters and surgeons) were expected to recommend actions in their area of expertise, not to promote operations pushing into the realm of line officers. But Benjamin Isherwood wasn't an ordinary staff officer—he pushed hard. And he was in no doubt that war was coming. In a late January letter to John Lenthall, he remarked "I hope the political squabble will soon blow over though I confidently expect a civil war sooner or later."[14]

Just two weeks after the new engineer-in-chief's influence first became felt at the Navy Department, *Merrimack* entered Secretary Welles' flow of correspondence as published in the *Official Records.* On 10 April he issued the following "confidential instructions" to Commodore McCauley:

> In view of the peculiar condition of the country, and of events that have already transpired, it becomes necessary that great vigilance

> should be exercised in guarding and protecting the public interests and property committed to your charge. It is therefore deemed important that the steamer *Merrimack* should be in condition to proceed to Philadelphia or to any other yard, should it be deemed necessary, or, in case of danger from unlawful attempts to take possession of her, that she may be placed beyond reach.
>
> Indeed, it is desirable that all the shipping and stores should be attended to, and should you think an additional force necessary, or that other precautions are required, you will immediately apprise the Department. In the meantime exercise your own judgement in discharging the responsibility that devolves on you.
>
> It is desirable that there should be no steps taken to give needless alarm, but it may be best to order most of the shipping to sea or other stations.
>
> Please keep the Department advised of the condition of affairs, and of any cause for apprehension, should any exist.[15]

The reader is immediately struck by the conditional language that Welles, a gifted writer, chose to employ in these most important instructions; he avoided issuing any specific orders. His first paragraph was keyed on the phrase "it is … deemed important" and the following two opened with "it is desirable." In suggesting to McCauley that "it may be best to order most of the shipping to sea," Welles seems to have entirely forgotten that his promised drafts of sailors never reached Norfolk. Exactly when McCauley received Welles' instructions is not known.[16] There is no record of a response by the commodore, a pattern that would become all too apparent as time went on.

Lurking deep in the secretary's mind, unknown to Isherwood or McCauley, was Welles' knowledge that a relief expedition was about to descend on Fort Sumter. His recent orders to *Powhatan*'s Captain Mercer specified that he arrive at Charleston bar on the morning of 11 April. Once Fox's expedition began relieving the fort, its message would be indisputable—Lincoln's passive policy toward the Confederacy had been abandoned. The outbreak of hostilities was then a very real possibility. In that event, the situation at Norfolk was hostage to sudden change. And if a confrontation off Charleston didn't propel Virginia into the lap of the secessionists, the news that *Merrimack* was about to be spirited out of Gosport surely would.

* * *

Following up on Welles' instructions, later the same day John Lenthall, chief of the Bureau of Construction, Equipment and Repairs and to whom Engineer-in-Chief Isherwood reported, sent a telegraphic order to McCauley "to prepare the *Merrimack* promptly for temporary service under steam alone," and asked to be advised of "the earliest date that she could be ready." The use of the telegraph meant Lenthall's query would immediately be public knowledge. This time Commodore McCauley responded the very next day, telegraphing back "Mr. Danby, chief engineer, says it will take one month to prepare *Merrimack*'s engines for the service contemplated."[17]

Danby's estimate was immediately seized on by Isherwood as being "so different from the private information received from the chief of the yard that misrepresentation was evident."[18] Had Danby's time frame been exaggerated by Southern sympathizers at Gosport? At this remove it's a difficult question to answer, but this much is known. By Isherwood's later admission, *Merrimack*'s "engines were in a wretched state. All the braces were out of the boilers ... and the entire machinery was in a disabled condition."[19] But other considerations may have colored Danby's or whomever else's estimate. While she was a steam frigate, *Merrimack*'s engines were always considered to be auxiliary to her sail power. Further, throughout her brief career, her machinery gave constant trouble and frequently broke down.[20] Now laid up at Gosport, she was "a dismantled hulk, with only her lower masts and bowsprit in, and without a particle of rigging."[21] Given Lenthall's demand that *Merrimack* be prepared for "service under steam alone," it is understandable how a chief engineer, reporting from a navy yard whose officers and workmen had exhibited secessionist leanings, might have offered a conservative estimate to ensure he had the time to address all the unknowns implicit in the recommissioning of an unreliable steam plant deep inside one of the Navy's largest warships while working in such volatile surroundings. Given the circumstances, it was not an entirely unreasonable reply.

But that was not how McCauley's response was interpreted by the Navy Department. It set off alarm bells that reverberated through the pages of the *Official Records*. If she had escaped attention before, the steam frigate *Merrimack* was now firmly in the Department's sights. At this crucial crossroads, Welles revealed his growing disquiet with Commodore McCauley:

> And in answer to inquiries which I made of him as to the time which would be necessary to put the engines or machinery of the *Merrimack* in order, so that she could be moved, he sent word that

> it would require at least a month. On receiving this answer, I became apprehensive that I could not depend on him if the emergency should demand prompt action.[22]

Thursday 11 April saw the unleashing of four orders by Secretary Welles, all of which dealt with the Gosport Navy Yard, three of them specifically addressing *Merrimack*. The first inserted a significant new dimension into McCauley's already growing burdens:

> Be pleased to have the *Plymouth* prepared immediately to be taken to Annapolis, Md., where she will be fitted up with special reference to the accommodation of the acting midshipmen, who will embark on the usual summer cruise.[23]

Plymouth, the naval academy's training ship, wasn't normally required for her seasonal duties until 1 June but was now ordered to be "immediately" available. The Academy's superintendent would write Welles four days later asking for her to serve as a guard ship.[24] Observe that the secretary made no mention of when her officers and crew would be dispatched. Welles' next three orders of the day dealt specifically with *Merrimack*:

> Be pleased to have the *Merrimack* prepared in as short a time as possible for temporary service. The commandant at New York has been directed to transfer to the receiving ship at Norfolk, without delay, 200 seamen, ordinary seamen, landsmen, firemen and coal heavers, for the purpose of making up a crew for the *Merrimack*. The Department desires you to inform it when it is probable the vessel will be ready.[25]

Welles dutifully sent the necessary order to the New York Navy Yard, concluding on a note of uncertainty. "If the requisite number are not on hand let them be enlisted as soon as possible."[26] It was the Department's first pledge to supply crews to McCauley since the attempt of 31 March had been thwarted by the Fort Sumter expedition. Why had they waited so long and when would the men arrive? *Germantown* was still unmanned and orders had now been issued to activate two more ships. Many more sailors than those already earmarked were needed if all three of Gosport's warships were to be adequately manned for their "temporary service."

* * *

Secretary Welles' final order of the day, actually two separate orders, was placed in the hands of Commander James Alden who was then in

Washington.[27] The first, intended for the file, ordered Alden to report to Commodore McCauley for temporary duty. The second order, Alden's "private instructions," read:

> You are directed to report to Commodore McCauley to take charge of the steam frigate *Merrimack* and deliver her over to the commanding officer of the Philadelphia station for the necessary repairs.[28]

Alden must have been told more than what can be read in the *Official Records*. His instructions, with its scheme to move *Merrimack* out of the Gosport Navy Yard, added a new kinetic element to the siege atmosphere that would soon envelop Norfolk. Writing in his diary and referring to this specific order, Welles later explained the reasons for his brevity:

> Many of the instructions in those days were given orally, for what became a matter of record was too often, in some mysterious way, made known to the insurgents. No more than was absolutely necessary was put upon paper for any of the officers who were sent to Norfolk.[29]

Appended to Alden's directive was a "memorandum" enclosing orders for four engineers, Danby and three others, who were assigned to accompany Alden on his mission. No line officers are mentioned and subsequent correspondence does not indicate any were assigned, a perplexing omission. At this stage, it is fair to ask who was James Alden, the man tapped by Secretary Welles to bring *Merrimack* to safety? History has indulged in a microscopic examination of Commodore McCauley and found him wanting, disparaging his person and reputation, but has spared Alden. Commander Alden was an experienced surveyor, served creditably during the Civil War and later retired as a rear admiral, but he was unsuited for the bold task Welles assigned him. At the time of his appointment by the secretary, James Alden, hailing from Maine, was a fifty-one-year-old commander waiting for orders. His previous twelve years had been occupied by Coast Survey duties, the last stint leading a party on the Pacific coast. Other than Coast Survey schooners and steamers, he had never commanded a large warship, certainly nothing the size of *Merrimack*, and his experience with steamers was limited to those of the Coast Survey, small side-wheelers. As far as age and infirmity, he only made the rank of commander at age forty-six. Compare that to Charles Stewart McCauley who had been promoted to captain at the same age. For many reasons—strength of character, leadership abilities and command experience—Commander Alden was not the right man for the job.[30]

* * *

Despite having successfully stirred the pot and motivating Secretary Welles to issue multiple orders aimed at removing *Merrimack* to safety, Engineer-in-Chief Isherwood was still not satisfied. "At his own urgent request" Benjamin Isherwood persuaded Welles to send him to Norfolk so that he could "in person, take full charge of the *Merrimack*, and get her ready as soon as possible."[31] The contrast between Isherwood and Alden is stark, one man at the very peak of the Navy's engineering hierarchy, the other an unknown Coast Survey officer. If *Merrimack* was worth sending the engineer-in-chief supported by a squad of assistant engineers, why wasn't a matching team of aggressively competent line officers dispatched? The Navy and *Merrimack* would suffer the consequences.

On Friday 12 April, Secretary Welles' writing machine kicked into action, and issued a pair of orders, the first marked "Confidential" to McCauley:

> The Department desires to have the *Merrimack* removed from the Norfolk to the Philadelphia navy yard with the utmost dispatch. The Engineer in Chief, B. F. Isherwood, has been ordered to report to you for the purpose of expediting the duty, and you will have his suggestions for that end carried promptly into effect.[32]

Welles' second order, addressed to Isherwood, simply instructed him to "Proceed to Norfolk, Va., and report to Commodore McCauley for such duty as he may direct."[33] Observe that an integral feature of Welles' sets of orders to both Alden and Isherwood was that while they outlined each officer's desired course of action, they left the ultimate discretion for each to act solely in the hands of the commodore. Both were to report to him and perform their duties "as he may direct." This feature would be the cause of later controversy and recriminations.

* * *

Unknown to Secretary Welles as he scribbled his signature on the dozen or so orders and instructions aimed at rescuing the stranded steam frigate *Merrimack*, was that one of the Virginia Convention commissioners appointed to determine President Lincoln's true intentions was just then strolling through Gosport Navy Yard, taking the measure of its extensive facilities. George Wythe Randolph, the commissioner favoring secession, had been laid over in Norfolk on his way to Washington, and following his return to Richmond gave an account to the Convention of what he had seen. Addressing a secret session, Randolph reported:

> Two things, Mr. President, ... are to be done quickly, and to some extent secretly. ... In the first place here is a dock-yard which it would be important for us to immediately possess. I took occasion to walk through it on Thursday [11 April] while in Norfolk, and to converse with the officer in charge of it, and to find out what they were doing. I ascertained that there are stored there 30,000 barrels of gunpowder, besides several hundred guns of heavy calibre. If you seize all that immense dock-yard, with all its ship timber, its vessels in the streams, your armament is complete. You have got what you most need; an ample supply of powder, vessels in stocks and materials for a navy force. If you stand idly by—even now they are stripping it of everything and moving it off—all that material will be employed to subjugate instead of defend you.
>
> There [are also] ... at Harpers Ferry 48,000 Minié muskets and a much larger number of smooth bore muskets. It would be an easy matter to prevent the removal of these ... The question now is whether you give up these arms to the enemy or keep them for yourselves.[34]

Randolph's speech makes it clear that the Virginians were keenly aware that plans were afoot to sneak valuable war matériel out of the Gosport Navy Yard and that Harpers Ferry was also a rich target. The choice was theirs—if the Old Dominion didn't first seize these military installations, their arms would be turned against her. As Randolph forcibly argued, "We do not need advice now. We need ... bayonets."[35]

Of course, Gideon Welles had been told nothing about Randolph's tour or that one of Gosport's officers, perhaps Commander Charles Robb or even McCauley himself, had had conversations with the visiting Virginia commissioner. It had been another busy week and he could hardly be excused for anticipating a restful weekend. As Welles confided to his diary in the wake of a similar hectic period, "I congratulated myself, when I went to my room at the Willard's on the evening," that so much had been accomplished within such a short time.[36]

But the secretary would get no rest. Only hours before, the opening shots of the Civil War had been fired.

* * *

CHAPTER 8

Fort Sumter Surrenders

With a perversity that only hindsight yields, the saga of the two hostage forts played out very nearly as Secretary Seward had desired. A belatedly reinforced Fort Pickens remained under Union control while Fort Sumter was grudgingly evacuated and handed over to Confederate forces. But these bald facts tell only part of the story. Fort Sumter's surrender came only in the wake of a fierce thirty-three-hour artillery barrage that battered it into submission, ushering in the American Civil War. And along the twisting path to this melancholy milestone, came the exposure of systemic failures at the very highest levels of the Lincoln administration.[1]

The seeds of the administration's missteps were planted in the very last days of March when Lincoln finally accepted the need to take decisive actions to save the forts. One prong, run through the War and Navy Departments, featured Fox's Fort Sumter expedition; the other, murkier scheme was concocted by Secretary of State William Seward. Secretary Seward conspired with Army Captain Meigs and Navy Lieutenant Porter to force a decisive action in Pensacola Bay intended to divert Confederate attention from the federal reinforcement of Fort Pickens. Unfortunately, their planning refused to acknowledge priorities already assigned to Fox's parallel effort. And in the earmarking of naval resources for their expedition, Seward hoodwinked President Lincoln into signing orders that bypassed Secretary Welles.[2]

The same day Welles changed course on *Powhatan*'s decommissioning, April 1st, he realized that Seward was up to something devious but just how far it reached escaped him. It was only five days later, hours after Welles had issued orders to *Powhatan*'s Captain Mercer to support Fox's relief effort, that the full extent became clear. And its scope was breathtaking. Secretary of State Seward had tricked President Lincoln into signing orders giving command of *Powhatan* to Lieutenant Porter in furtherance of

Seward's crusade to save Fort Pickens, and then hid them under the guise of sealed orders.[3]

This highly irregular protocol—the president issuing direct orders to the commandant of a navy yard and to the captain of a major warship with the intent to redirect that ship into the hands of a grasping lieutenant, all without notifying his secretary of the navy—came to light but not soon enough to prevent its disruption of Welles' finely balanced mobilizations in support of Fox.[4] The result was that *Powhatan* sailed as ordered on 6 April, but in command of Lieutenant Porter, not Captain Mercer, and shaped a course toward Pensacola, not Charleston. Before Welles could right the situation, she had disappeared over the horizon. And incredibly, although he was still in New York, Gustavus Fox had no idea she had been diverted.[5] Secretary Welles, writing later in his diary, didn't mince words in describing how this imbroglio affected the Fort Sumter operation:

> In detaching the *Powhatan* from the Sumter expedition, no important or necessary aid was furnished by her or by Lieutenant Porter to Pickens. ... No additional frigate like the *Powhatan* was needed there while she was indispensable here. That vessel gave no greater security to Pickens. ... Hence the reinforcements were stolen away from Sumter and sent to Pickens.[6]

Fox was delayed by stormy weather and his little armada only got to sea on 9 April, delaying his arrival off Charleston bar until early on the 12th. By then Confederate General Beauregard's gunners had opened fire on Fort Sumter. More time was lost by Fox waiting for *Powhatan*'s non-arrival and all the while heavy swells hampered small boat operations between his ships. Even worse, Fox's three chartered tugs, an essential element of his rescue plan, failed to arrive on time. Only one, *Yankee*, actually made it through the stormy seas to Charleston but was hopelessly late—more will be heard about this doughty steam tug.[7]

In the end bad weather, Southern gunners and Fort Sumter's inaccessibility thwarted Fox and his small force. Despite his zealous efforts, it was his fate to witness the shooting away of Major Anderson's flag. After an agonizing delay, it was raised again but soon replaced by the white flag of surrender. Two days later Fox embarked the fort's garrison for its return to New York, a sad ending for such a brave enterprise.[8]

The Lincoln administration had suffered a rude setback. Secretary Seward's power grab, the president's failure to detect his deceit, Porter and Meigs' connivance in the irregular diversion of scarce naval resources and Welles being thrown off balance during such a critical period caused a significant derangement of federal government function. Most directly it emasculated Fox's force and generated mistrust among members of Lincoln's cabinet. Even worse, it so averted Welles' attention from the deteriorating situation at Norfolk that his subsequent efforts to gain control of events were more reactive than proactive. His hurried recommissioning of *Powhatan* had already denied Norfolk its urgently needed drafts of seamen. Newly in office for only thirty days and faced with an unprecedented crisis—a sectional rebellion—the administration's executive ability to exert its will had been grievously compromised. That disruption materially contributed to the debacle that would soon befall the Gosport Navy Yard.

* * *

Virginia's three commissioners sent to sound out Lincoln found their progress mired by wild thunderstorms so severe that it took them four days to complete a journey that should have taken one. That put them in the nation's capital on 12 April, the same day Beauregard began his cannonade of Fort Sumter. Then another day passed before President Lincoln was free to see them.

Meanwhile, back in Richmond, their State Convention colleagues, who had now been making speech after speech for two full months, became increasingly aware of their waning importance. Waiting for their commissioners to report, unable to conclude any tangible business, they were no longer the focus of the Old Dominion's attention. With every day's headlines threatening imminent war and cheering on federal reinforcements, their eyes, like those of their fellow citizens, now gazed toward Washington.[9]

The night before the Virginia commissioners met with Lincoln, news of a fierce bombardment at Charleston reached the capital.[10] The next day, 13 April, as they began their audience with the president, Major Anderson surrendered Fort Sumter. The object of the commissioners' visit had now been fundamentally altered. Instead of determining Lincoln's policies toward the seceded states and Virginia, they now were witness to his stern warning: "If, as now appears to be true, … an unprovoked assault has been made upon Fort Sumter, I shall hold myself at liberty to repossess, if I can" any federal forts seized by the Confederacy.[11] The president's inaugural vow to "hold, occupy and possess" all federal property would be put to the test.

Lincoln's words marked a new firmness. He now stood ready to accept the risk of a war he had tried so hard to avoid, and in doing so emphatically rejected compromise with secessionists. Confederate artillery had propelled a handful of rebellious states' self-indulgent outburst into a higher realm, the preservation of the United States. Parsing constitutional issues over slavery was yesterday's business. The battle cry had been sounded. A divided and irresolute North quickly rallied behind the president's common cause—saving the Union. His emotional appeal was one that could, and did, grip all of its citizens.[12]

Two days after the commissioners' futile meeting, a bellicose proclamation by the president appeared on the front pages of all the papers—a call for 75,000 militia to act against "combinations too powerful to be suppressed by the ordinary course of judicial proceedings."[13] The specter of a military response now loomed. But even in its extremity of disunion, the federal government refused to name or define the "said combinations." In its desire to avoid legitimizing the Confederacy by naming it as such, and by not including Southern states' militia or active committees of safety among its foes, the federal government shrank from clearly identifying its enemy. "Combinations" literally meant "an instance of combining" with no implication of menace or threat beyond the vague suggestion of united action. Commanders of isolated military and naval installations needed definite instructions; it was not their duty to interpret a president's amorphous words. Would, for example, the state of Virginia be considered a "combination" of persons that should "disperse and retire peaceably to their respective abodes." It was hardly a moot point as events of the next week would so painfully prove.[14]

* * *

The force would be used to preserve the Union. That much was clear. But what resources were available to back up Lincoln? His call for 75,000 militia would greatly bolster the peacetime army's ranks of barely 17,000 officers and men, but time was needed to muster, outfit, equip and train the nation's newest soldiers. The Navy was in a similar bind. With only fifty ships in commission, any immediate response could be met only by bringing home frigates and sloops that were cruising on foreign stations, recommissioning vessels in ordinary and by the purchasing and converting of merchantmen. But each had severe limits. A good proportion of the Navy's men-of-war, both in commission and in ordinary, were sailing vessels which would be useless in enforcing the blockade of the Southern coast that would soon be declared. And converted merchant ships were at best an imperfect solution. A month would pass before John Lenthall, the chief of the Bureau of Construction, Equipment and Repairs, began to sketch out plans for the small steamers that gained the nickname "ninety-day gunboats," and more time would be lost before the first of these versatile warships could be built and placed in service.[15]

For now, Secretary Welles was forced to make do with what was immediately available. Until his orders to the far side of the globe brought back the Navy's cruisers or laid up vessels were reactivated, that meant the Home Squadron. One can spare a pang of sympathy for the secretary as he gazed at the *Navy Register*'s list of warships for that squadron—all twelve of them. What his predecessor optimistically described as "unusually large" now appeared wholly inadequate. As Welles knew, its flagship, *Cumberland*, was now standing guard at the Gosport Navy Yard, and the steam sloops *Pawnee* and *Pocahontas* were at sea in support of Fox's Fort Sumter expedition, backed up by the revenue cutter *Harriet Lane*. The State Department's demand for coverage off Veracruz was still being met by the sailing frigate *Macedonian*. The squadron's remaining eight warships were either standing by off Pensacola or, like *Powhatan*, headed there. Of these, three were sailing vessels and another three were small, third-class steamers typically mounting only five guns. Also available to the secretary in a pinch was the steam tender *Anacostia*, at the Washington Navy Yard, and the side-wheel gunboat *Water Witch*, fitting out at Philadelphia. And if he were really desperate, the sailing frigate *Constitution*, more a relic than a warship, was anchored off Annapolis as a practice ship.[16] That was it.

On that same eventful Monday that President Lincoln issued his proclamation calling up the militia, the 15th of April, Secretary Welles decided to replace certain of the officers at the Gosport Navy Yard. Perhaps he was inspired to do so by his recent dispatch of Commander Alden and Engineer Isherwood, and on the surface, his action had the appearance of an attempt to exchange officers hailing from Virginia with Northerners. Commander Robert Robb, the yard's executive officer and Lieutenant Charles Spotswood were issued orders detaching them on that day, Robb to "wait orders" and Spotswood to the command of the brig-of-war *Perry*.[17] Both were Virginians, but then Robb had been on duty at Norfolk for three years and Spotswood, rather than being sidelined was detailed to the command of one of the Navy's scarce warships. On the following day, Lieutenant John Watters, a Michigan native, was relieved of duty on the receiving ship *Pennsylvania* with orders to join the steam frigate *Minnesota*, and Lieutenant George T. Sinclair, a Virginian, was detached from his ordnance duties to "wait orders."[18]

Replacements for three of the four detached officers were men with Northern allegiances—Commander John Livingston of New York for Robb, Lieutenant Enoch Parrott of Maine for Spotswood, and Lieutenant John Irwin of Pennsylvania for Watters—thus creating strong the impression of a conscious effort to reduce the yard's Southern exposure. As for Sinclair, the details of his relief have not come to light. Closer examination of these exchanges suggests they were structured to create the façade of a routine reshuffling of officers, which may well have been the Navy Department's object given the scrutiny they would face at Gosport. Welles' diary was silent on this matter, leaving historians to draw their own conclusions. Whatever the intention, for good or for bad, these four sets of orders would amplify the loss of continuity within Commodore McCauley's team of officers at the worst possible moment.

Lieutenant John Irwin's assignment to Norfolk was tinged with irony: he would be a witness to history repeating itself. Irwin, the son of a former Pennsylvania congressman, had recently been on leave in Pensacola, attempting to move his family to his native Pittsburgh. But then Florida's secession and the threat of hostile militia reared its head. Temporarily detailed to the steamer *Wyandotte*, he participated in transporting the army's garrison to Fort Pickens and stood quietly at a distance, watching as the navy yard's flag was lowered. Irwin then requested that he be detached. Gathering up his family and fleeing to safety, he soon made his way north. And now Lieutenant Irwin had been issued orders to report to the Gosport Navy Yard.[19]

* * *

When the telegraph first brought news of the firing on Fort Sumter to the delegates of Virginia's State Convention, they discounted it as fake, just one more wild rumor designed to stir up a nation on edge. Putting their heads down, the delegates carried on with their focused debates. But former governor Henry Wise's reaction was very different. He vigorously urged his colleagues to put aside their talk and take action. But no, they pushed back with all manner of excuses as to why such rumors were of little consequence to them or the Old Dominion. The opening day of America's bloodiest war was passed by Virginia's delegates indulging in peaceful dialogue on the floor of the House of Delegates.[20]

While consumed with emotion, Henry Wise was also taken aback by President Lincoln's display of strength. He had expected the Yankees would evacuate the fort, and thinking in the mold of a Virginian statesman, that moves toward dissolution of the Union would only occur after serious debate and prolonged parliamentary maneuvers, not with a sudden thunder clap.[21]

But that is what came next and its effect was electrifying. Again, the news flowed to Richmond over the telegraph wires, spelling out the president's proclamation calling up the militia. Even worse, Secretary of War Cameron confirmed that the Old Dominion was expected to supply their share of 3,500 men. The powers in Washington had now demanded that Virginians join the fight against their fellow Southerners. [22]

Henry Wise quickly grasped what had transpired, indignantly declaring "I don't believe it."[23] The fog swirling around Lincoln's intentions had now been swept away. A turning point was fast approaching. After countless speeches, debates and finely balanced arguments, the Old Dominion's moment of truth had arrived. That day's *New York Times* figuratively threw down the gauntlet:

> Everybody here sees that now war has commenced, the question which the Virginia Convention has to decide is simply whether Virginia will declare war against the United States or stand by the Government; whether she will invite the battle upon her soil, to her utter ruin, or aid in bringing the fratricidal strife to a speedy termination by sustaining the Government and Union.[24]

But the Virginia delegates, huddled together in their Capitol building, each one of them keenly conscious of his higher moral duty, viewed the same events through a different prism. Henry Wise's grandson and biographer explained the emotions that energized them:

> The vast majority of the members of the Convention had been Union men, almost to the very last, but to the staunchest Whig, as well as Democrat, the conception of a Union meant a sisterhood of co-sovereign States, and not a consolidated government held together by force and arms. ... and the idea of the Federal Government, the creature of the several states and an agent of limited and derivative powers only, attempting to coerce and invade its creators, was repugnant to the teaching of her statesmen and the instincts of her people.[25]

Heated debates broke out among the delegates over the issue of whether to go into a secret session. Until this point in time, all of the Convention's proceedings were open to the public, reported on by hungry correspondents and eagerly devoured by the public. And there was no audience that more eagerly anticipated the latest news from the Convention than Gosport Navy Yard's cohort of Virginian naval officers, all of them alive to how their futures were hostage to what happened next in Richmond.

The news of President Lincoln's proclamation had thrown the Convention's Unionists off balance and set them at odds with the federal government even if the hostilities at Fort Sumter turned out to be a passing affair.[26] An address by Robert Scott, a pro-Union delegate from Fauquier County, made clear how far their position had shifted:

> So long as hostilities should be confined between the Administration at Washington and the Confederate States … I could see no imminent danger of a general war; but now it seems that the President of the United States has issued his proclamation making a requisition for 75,000 soldiers, … designed for no other purpose than active general war. … We have already declared—deliberately declared—that under no circumstances, in our opinion, ought the people of this State to consent to be made parties to an unnatural war against our Southern brethren. … but if … the administration is about to inaugurate a general war, … we must disconnect ourselves from this Federal Government.[27]

Although Scott had pleaded with his fellow delegates to remain in open session, the pressure to go into secret session steadily increased. As everyone knew, once that happened the talk would turn to Virginia's secession. And no one could argue that case with more elegance than the man impatiently waiting to speak, former governor Henry Wise.

* * *

CHAPTER 9

Merrimack—The Prize

Both Commander James Alden and Engineer-in-Chief Benjamin Isherwood had been ordered at short notice from Washington, D.C. to important duty in Norfolk—saving the steam frigate *Merrimack*—Alden on 11 April and Isherwood the next day. Their later correspondence is silent about when Alden arrived, revealing only that Isherwood entered the Gosport Navy Yard on the morning of Sunday the 14th. It is likely that Alden was already there by then. So how did they get there? The next week in the life of the Gosport Navy Yard featured an increasing number of sudden visits and departures, nearly all by officers shuttling between the nation's capital and Norfolk. The mechanics of their travel—the schedules, duration of trips and the routes—exerted a subtle influence on events, so it is a germane question. They traveled to Norfolk courtesy of the Baltimore Steam Packet Company, or as it was commonly called, the Bay Line.

Formed in 1840, the Bay Line ran a fleet of steamboats between ports principally within the Chesapeake Bay and its tributary rivers. The core of its service was the 145-mile run between Baltimore and Old Point Comfort, the landing for Fortress Monroe, and then on to Portsmouth, some fifteen miles away, with the final connection to Norfolk via a local ferry. Direct service from Washington to Old Point Comfort was also scheduled, but as that run featured mainly passengers it was not as frequent as the more heavily subscribed passenger and freight service offered from Baltimore to points south.[1]

The Bay Line was renowned for its luxurious service. Its steamboats' staterooms featured comfortable berths, their dining salons were fitted with imported Belgian carpets, velvet-covered chairs and marble-topped tables, and their public spaces were set off by white paneling with gilt moldings. The line's popular overnight service was keyed to a late afternoon departure, 5 p.m. from Baltimore, with arrival at Old Point Comfort the next morning at 6 a.m. After a brief stop, Portsmouth was reached a little over an hour later. A mirror-image schedule facilitated returns from Norfolk to Baltimore. Service to Washington was also shaped around over-night passages.

Large side-wheel steamboats were employed on the Bay Line's routes. To meet the line's demanding timetables, they could maintain speeds of twelve knots. In 1859, the first iron-hulled steamer was introduced, *Georgeanna*.[2] With their walking beams pumping up and down, frothy wakes spreading wide from their fast-turning paddle wheels and a whistle's bray to mark their passage, the line's steamboats must have presented a magnificent spectacle charging up and down the Chesapeake Bay.

* * *

On the same Sunday morning that Engineer-in-Chief Isherwood recorded his arrival at the Gosport Navy Yard, the foreign correspondent for the London *Times*, William Howard Russell, was completing his passage to Norfolk on *Georgeanna*, very likely the same boat Isherwood traveled on. Russell's stories fashioned from his early Civil War reporting, *My Diary, North and South*, sketch a vivid portrait of her voyage's end at Portsmouth:

> The steamer in a few minutes came along-side a dirty, broken-down, wooden quay, lined with open booths, on which a small crowd, mostly of negroes, had gathered. Behind the shed there rose tiled and shingled roofs of mean dingy houses, and we could catch glimpses of the line of poor streets, narrow, crooked, ill-paved, surmounted by a few church-steeples, and the large sprawling advertisement-boards of the tobacco-stores and oyster-sellers, which was all we could see of Portsmouth or Gosport. Our vessel was in a narrow creek; at one side was the town—in the centre of the stream the old "Pennsylvania," intended to be of 120 guns, but never commissioned and used as a receiving ship, was anchored—alongside the wall of the Navy Yard below us, lay the "Merrimac," apparently in ordinary. The only man-of-war fit for sea was a curiosity, a Dutch-built looking sloop, called the "Cumberland." Two or three smaller vessels, dismasted, were below the "Merrimac," and we could just see the building-sheds in which were one or two others, I believe, on the stocks. A fleet of oyster-boats anchored, or in sailless observance of the Sunday, dotted the waters. There was an ancient and fishlike smell about the town worthy of its appearance and of its functions as a seaport.[3]

Russell's descriptions of Tidewater Virginia are uniformly demeaning so the reader is cautioned to apply a filter. Russell spent a night at the same

Norfolk hotel in which Isherwood stayed, and his picture of this famous establishment was equally grim:

> An execrable, tooth-cracking drive ended at last in front of the Atlantic Hotel, where I was doomed to take up my quarters. It is a dilapidated, uncleanly place, with tobacco-stained floor, full of flies and strong odors. The waiters were all slaves: untidy, slipshod, and careless creatures.[4]

And finally, before leaving Russell's first-hand sketches, later that day he encountered a boat's crew from the sloop-of-war *Cumberland*, awaiting their coxswain's errand ashore and who had just been pelted by oyster-throwing, obscenity-shouting locals. News of Fort Sumter's demise was the talk of the town and *Cumberland*'s men represented the despised Yankees:

> "Who threw the shells?" No one answered at first, but a little wizened man at last squeaked out: "I guess you'll have shells of another kind if you remain here much longer." The sailor howled with rage: "Why, you poor devils, I'd whip any half-dozen of you—teeth, knives, and all—in five minutes; and my boys there in the boat would clear your whole town. What do you mean by barking at the Stars and Stripes? Do you see that ship?" he shouted pointing toward the "Cumberland." "Why the lads aboard her would knock every darned seceder in your State into a cocked hat in a brace of shakes!"[5]

At this moment in time the mounting tensions between North and South were already animating the citizens of Norfolk. Alden and Isherwood's experiences attempting to recover *Merrimack* will, if anything, even more strongly reinforce Russell's caricatures.

* * *

Commander Alden's account of his arrival at Norfolk tells that the locals had "already placed obstacles in the river and set on foot other warlike preparations." Alden's "river" is, of course, the Elizabeth River, and the obstacles' location was close by Craney Island, where the Confederate "stars and bars" were now flying over the old War of 1812 blockhouse.[6]

The spies that hounded Welles' Navy Department had been hard at work, broadcasting to all that the commander had come to take out *Merrimack*. Alden recorded that:

> all Norfolk seemed to be full of it when I arrived there, and they had made up their minds to stop the ship … Finding my movements in the city were watched so closely, and the attitude of the people toward me so threatening, I destroyed my private instructions and took up my abode at the navy yard to wait for the ship.[7]

Isherwood, in contrast, went straight at it. First locating Engineer Danby, he then presented his orders to Commodore McCauley together with a copy of Welles' instructions of 12 April, expressing the Department's desire to remove *Merrimack* from Norfolk. McCauley told Isherwood to do whatever was needed, and the two engineers got to work.[8]

Their job was colossal. Isherwood later described the machinery's condition as "wretched." All the braces to the frigate's four boilers had been removed in preparation to replacing them, the engines were completely dismantled and their parts scattered among the different workshops, but "nothing … was in such bad condition as to prevent its temporary use." Temporary was the key word—the two engineers had an even bigger job than they let on. *Merrimack*'s machinery had already been condemned because of chronic design faults: undersized bearings; inadequately dimensioned valves, pipes and cocks; weak linkages; pistons that were too big and faulty air pumps. Her engines had given trouble soon after she was commissioned; now after years of hard pounding, temporary fixes and running repairs, they were in worse shape than ever before.[9] Isherwood's task was beyond simply recommissioning a disassembled engine, it was to breathe a few day's life into a thoroughly defective set of machinery.

Undaunted, both Isherwood and Danby bent to their work with a will. Leveraging their high personal esteem among the yard's boiler-makers and machinists, they managed to quickly recruit a workforce large enough to stand three eight-hour shifts per day and supervised their work by handing them off every twelve hours, gaining further support from their trio of assistant engineers. That they were able to do so opened a window into another layer of Gosport's crises of loyalty. The yard had originally employed up to 1,500 workmen, nearly all locally hired, but the current state of unrest had slowed down work such that by late April, only 950 men were drawing wages. Curiously, the town of Portsmouth had elected two Unionist delegates to represent them at the State Convention, both by a large majority over their secessionist rivals. That *Merrimack*'s engines were coming together with every passing hour was evidence that a core of Virginians still supported a united America. And not one of them wore a blue cockade.[10]

All the while Isherwood was supervising his workers, he was also taking the measure of *Merrimack*'s needs in anticipation of her removal from Norfolk. He managed to load on board enough coal together with the necessary engineers' stores to get her out of the Elizabeth River and had somehow talked forty-four men from outside the yard into serving temporarily as firemen and coal-heavers. Again, that he was able to recruit a black gang given the local secessionist sentiment was astonishing. He even identified some with enough experience at sea to assign as helmsmen to steer the ship. Still, even with lavish promises of money, they refused to take her any further than Newport News on the James River's north shore.[11]

Meanwhile, Commander Alden was presumably preparing the big steam frigate for sea but of his efforts, we know next to nothing. During this period *Merrimack* was tied up alongside the yard's wharf, down from the ship houses. What accounts fail to emphasize when discussing the attempt to remove *Merrimack* from Gosport was her sheer size. With the exception of her four decommissioned sisters, she was the Navy's largest ship, bigger in terms of length and tonnage than the mighty 120-gun *Pennsylvania* lying just across the stream. At just over 256 feet long on the waterline and displacing 4,635 tons at full load, she positively dwarfed the frigate *Constitution* still afloat today in Boston. Although a valuable prize for either North or South, during those hectic April days *Merrimack* hardly resembled the dashing man-of-war that turned heads four years before during her inaugural cruise to Europe. Save for her lower masts and bowsprit, she had been shorn of all her upper masting, spars, sails and rigging. Her gun decks were empty as were her holds, storerooms and lockers. The officers' accommodations were stripped of furniture, and her sailors' hammocks were stowed ashore. Except for the coal and water Isherwood had taken on board, her bunkers and water tanks were empty. All her small boats had been removed. Sitting high in the water, *Merrimack*'s draft was not less than twenty feet, exposing the tips of her propeller. Her forbidding black topsides towered nearly thirty feet in the air.[12]

Getting a dead ship this massive ready for sea, even for a short voyage across Hampton Roads, took considerable preparation. With the exception of donkey engines to assist with loading coal and to power her bilge pumps, she had no steam-powered conveniences to help her crew. Anchors were cast and raised solely by manpower; steering was by hand via a large double wheel and all line handling was manual. Without her usual fore staysail and aft spanker to control yawing, the windage from her high sides would have made her extremely difficult to handle in the close waters of the Elizabeth River. Even getting her off the wharf and headed down river would take the efforts of skilled crews to "wind" the ship, or turn her around in the stream.

What is known about her condition prior to her intended departure is that she had no guns, no sails or rigging, no crew or assigned officers, had not been winded and only a few days of coal and stores were on board.[13] Compare that with the thorough preparations Engineer-in-Chief Isherwood made in the same timeframe. Why was he successful and Alden a failure? Alden's inexperience commanding a ship of this size may have been a factor, but the root cause can almost certainly be attributed to the paralyzing effects of the yard's Southern sympathizing officers. Gosport's executive officer, Robert Robb, and ordnance officer, John Tucker, both commanders, and the yard's first lieutenant were all known to have slowed Alden down at a later stage.[14] They were line officers and their cumulative drag would have most directly stymied Commander Alden's efforts to prepare *Merrimack* for sea, operations they well understood. None of McCauley's officers were engineers, so Isherwood and his team—lowly staff officers—were pretty much left alone to get on with their grimy work. In the secessionist-tainted atmosphere of the Gosport Navy Yard, Isherwood thrived while Alden wilted.

* * *

Isherwood's progress with *Merrimack*'s machinery prompted Commodore McCauley to issue one of his rare progress reports to Secretary Welles. In fact, it was the last one he would send before the yard was lost. When later questioned by the Senate committee about his dearth of representations to the Department, implying he should have written more often, McCauley replied "Certainly, as things have turned out, it would have been better had I done so, but as I knew the department was perfectly cognizant of my position ..."[15] Well knowing the Navy's concern over *Merrimack*, he now rose to the occasion. On Tuesday, 16 April McCauley advised Welles:

> I have the honor to inform the Department that the steam frigate *Merrimack* may now be taken and used for temporary service as soon as the necessary equipments can be put on board. All that is required to be done to the hull for temporary service will probably be completed by to-morrow evening.[16]

Note the conditionality of McCauley's statement: *Merrimack*'s readiness hinged on getting the necessary equipment on board. While McCauley's message made its way north to distant Washington, Secretary Welles was preparing to take the Department's response to a new, more muscular level. At some point, Welles must have given thought to the idea of removing McCauley—faithful but lacking the vital spark of initiative—and replacing him with an energetically inclined commandant. But because he had served only one-third of his term of duty at the Gosport Navy Yard, getting rid of McCauley was not a simple matter:

> To remove him would have necessitated extensive changes, involving an entire reorganization of the government of that yard, and consequently a departure from the President's policy of permitting things to continue undisturbed in Virginia. [17]

If Welles was constrained from replacing McCauley, the next best thing he could do was to dispatch a senior officer to meet with the commodore, assess his performance and based on the situation at the yard, especially the reliability of the officers and men, urge any necessary actions, much as Captain Levin Powell had done for Secretary Toucey. Welles decided for good measure to invest his chosen messenger with plenary powers to act on his behalf if danger was imminent. The man he selected for this delicate but momentous mission was a sixty-three-year-old captain attached to the Department's Office of Detail named Hiram Paulding.[18]

A native of New York, Paulding's early career echoed McCauley's youthful experiences in the Navy. As a midshipman, he served with distinction at the Battle of Lake Champlain, and later gained sea time with most of the Navy's foreign squadrons, including command of the 44-gun frigate *St. Lawrence* in 1850 on the Mediterranean. Paulding also served as the commandant of the Washington Navy Yard from 1853 to 1855, and soon after was assigned to be the flag officer of the Home Squadron. Here his instincts got ahead of good judgment when he was tasked with quashing further civil unrest in Nicaragua at the hands of the American filibuster William Walker.[19] Flying his flag from the steam frigate *Wabash*, *Merrimack*'s sister, Paulding took it upon himself to use the threat of imminent bombardment to seize and forcibly remove Walker and his armed followers from the Nicaraguan port of San Juan del Norte. Flag Officer Paulding's flagrant breach of sovereignty was alternately greeted by the American public with howls of condemnation and vigorous approval. In the end, Secretary Toucey was forced to mildly reprimand Paulding and suspend him from active duty.[20] Contrast Paulding's treatment following his command of the Home Squadron with that of McCauley's; one was abruptly retired and the other was given a dinner by the president.

The change of administrations brought Paulding back to favor and with hostilities in the offing, he found himself assigned to special duty at the highly visible Office of Detail.[21] Now in the Washington limelight, Captain

Paulding soon came to the notice of Secretary Welles. But not all of Paulding's contemporaries were enamored of his appointment and abilities; consider this portrait by his fellow officer, Captain Charles Wilkes:

> One of the great mistakes the Secretary made was to establish an officer, Comdr [Hiram] Paulding, in the Bureau for the assignment of officers. He was entirely unfitted for such duty, with strong prejudices against Southern officers ... He was unfitted in many other respects, ignorant of his profession and dicta[to]rial in his manner.[22]

For better or for worse, this was the messenger chosen to carry a batch of new orders to Norfolk. He was also expected to explain the views of the Department to both Commodores McCauley and Pendergrast, now under pressure in the increasingly isolated fastness of the Gosport Navy Yard.

* * *

The full text of Welles' orders to McCauley and Pendergrast, dated 16 April 1861, may be found in Appendix 3. So what were their principal elements and did they shed any new light on how Norfolk fit within the Navy Department's near-term priorities?

First, Secretary Welles' letter to Commodore McCauley. Welles' opening two sentences both referred back to his earlier communication of the 10th and, alluding to "the events which have transpired" since then, he suggested they demanded "additional vigilance and care in protecting the public property under your charge, and placing the vessels and stores, if necessary, beyond jeopardy." But Welles avoided identifying an explicit threat or even mentioning the fall of Fort Sumter or Lincoln's mobilization of the militia, stories of which were filling the headlines. His letter did, however, spell out several specific instructions and provided some news for McCauley's attention:

- Welles informed the commodore that Isherwood reported *Merrimack* would be able to depart on Thursday (the 18th), but offered his opinion that "It may not be necessary, however, that she should leave at that time, unless there is immediate danger pending." However, Welles stressed that "no time should be lost" loading her armament and valuable public property on board.
- Turning to *Plymouth* and *Dolphin*, Welles ordered that they should "be placed beyond danger of immediate assault at once, if possible." And *Germantown*, after being

loaded with "stores and ordnance from the yard," could be towed out by *Merrimack* "if an assault is threatened." But the only way to remove *Plymouth* and *Dolphin* from "beyond danger of immediate assault" was by sailing them out of Virginian waters. And how could *Merrimack* tow *Germantown* if she wasn't properly manned? For these operations, crews and their officers were needed the status of which was Welles' next subject.

- As for the promised drafts of sailors to be sent from New York, Welles informed McCauley that "recent demands have left an insufficient number to meet the requisition." For the second time in less than a month, the extra crews, so fundamental to moving manpower-intensive sailing warships, would not be forthcoming. Lincoln's higher priorities of Forts Sumter and Pickens prevailed. And Welles made no mention of detailing extra officers. McCauley was told to look to Pendergrast and *Cumberland* for any needed officers and men.
- Welles next introduced Commodore Paulding as the bearer of his orders and instructed McCauley to "advise with him freely and fully as to your duties and the interests of the Government in the present threatening emergency." Copies of Welles' letters, both of 10 and 16 April, were to be submitted to Commodore Pendergrast "who will assist and cooperate with you in carrying the views of the Department into effect."
- Secretary Welles concluded his communication by stating: "The vessels and stores under your charge you will defend at any hazard, repelling by force, if necessary, any and all attempts to seize them, whether by mob violence, organized effort, or by any assumed authority."[23]

Second, Secretary Welles' letter to Commodore Pendergrast, the full text of which also appears in Appendix 3. Welles' instructions were, thankfully, less demanding of the idled Home Squadron's flag officer. He was told that regarding the Department's plans for his flagship to return to Vera Cruz, "A state of things has risen which renders the immediate departure of the *Cumberland* as originally intended inexpedient." The "insufficiency of men in that station" required Pendergrast to "render assistance from the force under your command" to reinforce Norfolk and help McCauley. As for his cooperation with Paulding and McCauley, Pendergrast was enjoined to "advise freely and fully with both these gentlemen." Amplifying his first letter's instructions to McCauley, Welles further emphasized that Pendergrast was expected to "lend your assistance, and that of your command, toward getting the vessels now in the yard in condition to be moved." Welles concluded with a generic admonition that "in case of invasion, insurrection, or violence of any kind to suppress it, repelling force by force if necessary."[24]

Third, Welles' orders to Commodore Paulding. The official version was brief and to the point; the private version was undoubtedly more substantial. After instructing Paulding to proceed to Norfolk without delay, he was told "Upon completion of this duty you will return to Washington City and resume your regular duties."[25]

The Secretary of the Navy had just signed orders for two captains acting as commodores and instructed a third captain to deliver them who was also expected to explain the Department's thinking. That man, Hiram Paulding, ranked sixteen on the captain's list, McCauley ranked seven and Pendergrast, further down at thirty-one.[26] A Navy captain was always very conscious of his rank. And triggered by Paulding's appearance, Commodore McCauley might be expected to recall Paulding's recent reprimand as flag officer of the Home Squadron and compare it to his own widely praised performance. McCauley would no doubt carefully listen to Paulding and closely scrutinize the orders he carried, but being the senior officer, he would likely keep his own counsel. He alone was the yard's commandant. As for the flag officer of the Home Squadron, Commodore Pendergrast would well understand that he had been ordered to assist McCauley and that he was not expected to take the initiative. And his ugly confrontation with McCauley over orders to move *Cumberland* must not be forgotten.

Beyond their sensitivity over rank, it should be remembered that these three officers had grown old in a navy that had largely been at peace for the past fifty years. They were all only gradually promoted into positions of individual responsibility. During the greater part of their careers, they had been cocooned in a world where they did largely what someone else ordered them to do. The dash and initiative nurtured by the crucible of war were long gone. And the perplexities of a civil insurrection fomented by their fellow countrymen which pitted states against each other were beyond their experience. At best they could be expected to doggedly carry out their orders, attempting to reconcile any contradictions they posed, taking refuge in their time-worn concepts of duty.[27]

Welles' packet of orders tucked safely away, on the evening of Tuesday, 16 April, Commodore Hiram Paulding rushed to catch the Bay Line's boat to

Norfolk. He may have anticipated his special duty would be quickly and quietly concluded, like Captain Powell's had been three months before. Little did he realize how his mission's scope would expand beyond all recognition.

* * *

CHAPTER 10

Virginia's Agony, Wise's Insurrection

Lincoln's militia call-up left Richmond stunned and angry. The pressure for Virginia's Convention to go into secret session grew with each hour. Once that corner had been turned, open talk of secession could begin. All the delegates knew they were bound by a pledge to submit whatever decision the Convention made to a popular referendum, but alternately, it could be argued that preserving the people's welfare when the threat of war loomed justified the suspension of normal democratic protocols. Henry Wise's unique contribution to the Convention was his enunciation of this logic, *pro salute populi*, or "for the safety of the people." In these turbulent times, extraordinary measures were warranted:[1]

> Gentlemen do not seem to apprehend the correct idea of secret session. Men ... go into secret session because ... there is a secret in the subject itself. I ask the gentlemen if any subjects are fuller of secrets—of mysteries upon mysteries—than the subject of war and peace? ... Have you not to consider steps and preparations *pro salute populi*? Are you, guardians of the people, called to look over their safety, seeing them about to be destroyed, to wait to ask the people where the preparation shall be made to save them?
>
> Oh! The sweet mysteries of peace—Oh! The dreadful mysteries of war! Sir, for God's sake quit talking ... The time for action has now come. Action, sir, with lips not only unopened but compressed to immovable firmness ... What Virginia will do? Sir, with a view to answer that question quietly, deliberately, we are asked to go into secret session.[2]

In response to Wise's impassioned plea, John Baldwin, who had met privately with Lincoln only two weeks before—now it seemed like an eternity—moved to adjourn the session while the Convention waited on a fuller understanding of Lincoln's call-up of the militia. Governor Letcher dutifully sought clarifications and once received, the awful truth propelled the Convention into secret session. As of Tuesday, 16 April, the same day Commodore Paulding left for Norfolk carrying Secretary Welles' packet of orders, Virginia's Convention delegates locked their doors to the public's gaze.[3]

From without the populace was increasingly agitated. A Spontaneous People's Convention had been convened nearby and fears of impending anarchy in Virginia's eastern counties abounded. Seemingly unfazed by the threats of violence, the State Convention droned on, debating proposals for a fresh state-wide vote that would offer the options of a conditional secession or a conference of border states. Adding fresh fuel to the fire, Governor Letcher received two telegrams from a "trusted and respectable source in Norfolk," advising that the evacuation of the Gosport Navy Yard had begun and requesting authority to obstruct the Navy's retreat. No doubt Isherwood's progress repairing *Merrimack*'s engines had captured the insurrectionist's attention. The governor quietly and without endorsement laid the telegrams before the Convention.[4]

The delegates' speeches had already touched on the seizure of both the navy yard and the arsenal at Harpers Ferry, the secessionists arguing the case for action before it was too late and the Unionists cautioning against such a flagrantly war-like undertaking. That these two valuable federal properties, together with their munitions, arms, military and naval stores not to mention Gosport's fleet of warships, were in the Convention's crosshairs was undeniable.[5] The vulnerability of the federal position in Norfolk was also very visible to a local Union sympathizer, who wrote that day to a Northern correspondent:

> We are imperiled here. ... Our river channel is to be blocked just below the navy yard, to prevent any more Government vessels leaving. Old hulks have been bought for the purpose. There is no Government force here to prevent it. Virginia will secede ... I have no question of it.[6]

* * *

On this first day of the secret session, a tired and dispirited Henry Wise walked away from a committee meeting and alone in his thoughts, strode down Tenth Street near Richmond's Capitol Square. His Committee on Federal Relations had just rejected another of his proposals for Virginia to adopt a stand of armed neutrality, and he wondered if the state would ever be capable of taking decisive action. He suddenly brightened at spotting a familiar face, John D. Imboden, a thirty-eight-year-old lawyer hailing from the Valley. Wise well recalled Imboden had been defeated as a Convention candidate—he lost to a Unionist—and that he was the captain in command of the Staunton Artillery, a volunteer company of militia.[7]

After exchanging pleasantries, Wise quickly changed the subject. "Do you remember, sir, what passed between you and me, when I was governor, at the moment when you thanked me for the order permitting you to have two brass field pieces for your company of artillery?" Imboden replied, "Yes, I do, sir," and told Wise that he was duty-bound to obey his call for those guns whenever made. Henry Wise responded, "What was joke then, is in earnest now. I want those guns with which to aid in the immediate capture of the United States Arsenal at Harpers Ferry; can they be had with all the men you can raise?" Captain Imboden replied, "They can, and if you say so, the men shall be raised and the arsenal taken."[8]

Henry Wise's private insurrection had begun. Both men were excited by the prospect, Imboden to assuage his lost seat at the Convention and Wise to force Virginia's hand. At Wise's direction, Imboden hastily summoned six "reliable and brave" volunteer company commanders to a meeting at seven that night in Richmond's Exchange Hotel. Their first action was to dispatch a committee of three to extract a promise from Governor Letcher that, contingent on the Convention passing an ordinance of secession, he would telegraph the order authorizing their volunteer artillery and cavalry companies to move on Harpers Ferry.[9] The committee quickly gained Letcher's consent and returned to the hotel. Wise then faced his conspirators and said, "Well, gentlemen, you have heard the report; are you willing to act on your own responsibility?" They unanimously voted to accept Wise's challenge, acknowledging him as their de facto governor; then Wise began furiously scribbling the necessary written orders. By 11 p.m. the operation had been mobilized and the meeting adjourned.[10]

It was only when Wise was leaving the hotel, weary but exhilarated, that a clerk handed him a dispatch from William H. Parker, an acquaintance of his from Norfolk. It read: "The powder magazine here can be taken, and the Yankee vessels can be captured and sunk, so as to obstruct the harbor. Shall we do it?" Without hesitation, Wise immediately telegraphed back "Yes."[11] Parker was a member of a vigilance committee, which had long been promoting secessionist control of Norfolk and its environs. He or one of his colleagues may have been the source of the pair of telegrams sent earlier that day to Governor Letcher.

* * *

Engineer-in-chief Benjamin Isherwood was a silent witness to the Norfolk side of the conspiracy to block the Elizabeth River. In his later testimony to the Senate committee investigating Gosport's loss, he recalled events that occurred when he was overseeing the restoration of *Merrimack*'s engines:

> During the time I was at Norfolk I lodged at the Atlantic Hotel, the proprietor of which, Mr. Newton, was one of a committee in correspondence with Governor Letcher, who had directed them, in advance of the passage of the ordinance of secession by Virginia, to obstruct the channel and prevent the egress of the national vessels. I learned this by accidentally overhearing a conversation between Mr. Newton and another of the committee. The committee were then engaged in procuring small vessels to be filled with stone and sunk in the channel; but up to the time of my departure only two light-vessels had been sunk. They were about eighty tons burden, and their masts projected above the water, distinctly pointing out their location. … The committee were very fearful of obstructing the channel lest the "Cumberland" should turn her batteries upon the cities of Norfolk and Portsmouth, and were clearly desirous to be rid of the vessels without the perils of a collision.[12]

What actually happened on that night of 16 April was shrouded from eyewitnesses by weather "very dark and rainy, so thick that you couldn't see a boat's length," but the accounts agree on the main points. At 11:30 p.m. a tug was sighted from *Cumberland*, then lying at the naval anchorage off the hospital. The tug was steaming down the Elizabeth River with two light-boats in tow. Note that the time of *Cumberland*'s sighting aligned precisely with Henry Wise's telegraphed approval to "obstruct the harbor." Thirty minutes later a second tug was seen, also towing a light-boat down the river. These three light-boats, all previously condemned and lying in Norfolk's harbor, were apparently seized by the conspirators, filled with stone and once they were towed into position, scuttled in the narrowest part of the channel between Sewell's Point and Craney Island. While Isherwood reported that only two boats had been sunk, it appears a third light-boat was also scuttled. Open boasts were also made at this time that "*Cumberland* should never be permitted to leave the harbor."[13]

At the most basic level, it would seem that the obstruction of the channel was meant to prevent the escape of U.S. Navy vessels. But was it that simple? The conversations Isherwood overheard contained the strong suggestion that a psychological element drove the secessionists' actions. The Southern-leaning author of a history of the Confederate States Navy, Thomas Scharf, further fleshed out that theme: far from attempting to bottle up Norfolk's harbor, the obstructions were "part of the intimidation mode of assault" conceived to exploit the Navy's very real fear of entrapment. In his words, the scuttling of the light-boats was "designed only to threaten the closing of the river" and that a gap was purposely left between the sunken hulks, "a golden bridge for a flying foe," to permit the federals' escape. Isherwood later pointed out that he didn't think the light-boats presented an obstacle to *Merrimack*'s departure. In short, the visible and obvious placement of the obstructions was intended to scare the U.S. Navy out of Norfolk, the sooner the better.[14]

Such tactics were a classic weapon of a weaker opponent attempting to coerce its stronger foe into taking a rash action. But the secessionists were playing a dangerous game. Were the officers of the U.S. Navy expected to grasp such subtleties? Sinking a ship is not an exact science, especially if done in the dead of night and with impaired visibility. Even if a vessel is accurately located before submerging, currents and tides, contours of the seabed and scour effects greatly complicate any efforts to precisely position the sunken hulk. The simple fact was that obstructions had been placed in the Elizabeth River and more could be expected with each passing day. That was the message heard by Commodore McCauley.

* * *

Apart from obstructions to the Elizabeth River and their relentless intimidation, what other forces were available to the Norfolk area secessionists? Fortunately for the U.S. Navy and its interests in the region, the insurgents had absolutely no forces afloat beyond what they could beg, borrow or steal. However, on land local volunteer militias had long been part of the communities' defenses.

Closest to the Gosport Navy Yard, Portsmouth was host to five volunteer companies among which were the Portsmouth Light Artillery, which proudly drilled with its four guns. Across the river in Norfolk were another five companies including the Light Artillery Blues, also equipped with four guns. Two other county militia formations boosted their numbers, making for a total of twelve companies which could, in April 1861, field some 850 men. To them must be added the only cavalry outfit in the area, the Wise Light Dragoons. And while these companies were armed they had no ready supply of ammunition. Looking further afield, the state of Virginia was home to many other artillery companies, including Captain John Imboden's, but none were equipped with anything heavier than 12-pounder howitzers.[15]

Of the quantity and quality of the local and state militia that might be turned against his navy yard, Commodore McCauley had but scant information. Although he had "received a great many anonymous letters" threatening him and Gosport, his only source of information about outside forces came from people whom he mistakenly thought he could trust, citizens of Portsmouth. Perhaps McCauley, who had only been on the ground at Gosport since August last, could be excused for not having a greater knowledge of the local and state forces. But the same should not have been true of the man whose flagship had been ordered to Norfolk to protect the Gosport Navy Yard, Flag Officer Pendergrast, and whose wife lived in nearby Norfolk. Pendergrast testified that up until the time when the channel was first obstructed, he was "ashore daily," no doubt active in his wife's social world. She was a Virginia Barron. Yet when asked about the accuracy of his information regarding the forces outside the yard, he answered "It was not accurate because I had no opportunity of knowing personally about it."[16]

One of the most fundamental tenets of military science is to know your enemy. As the state of Virginia lurched toward disunion, the fate of the Gosport Navy Yard rested in the hands of a pair of commodores who had seemingly forgotten that basic rule.

* * *

IMAGES

NB: The images of three U.S. Navy officers—James Alden, Thomas O. Selfridge, Jr. and Hiram Paulding—show them at later stages in their careers with higher ranks than they held during the period of Gosport Navy Yard's destruction.

No. 1 *Commodore Charles Stewart McCauley*

This line engraving of Commodore McCauley gives the basics of his appearance: side whiskers, a firm jaw and a prominent nose that set off a care-worn countenance. Sixty-seven when appointed commandant of the Gosport Navy Yard, McCauley bore a passing resemblance to his uncle, Captain Charles Stewart of *Constitution* fame (From *Lamb's Biographical Dictionary of the United States,* Vol. V, 195).

No. 2 *The Navy-Yard at Norfolk, Virginia*

Peace before the storm. Appearing in the 16 March 1861 issue of *Harper's Weekly,* this image presents a view of the Gosport Navy Yard from the Elizabeth River. Dominating the left foreground is the receiving ship *Pennsylvania.* In the background are the two huge ship houses that flank the heavy lift sheers on the wharf, and to the right is the yard's northern-most row of buildings which enclosed its main gate (Photo # NH 42714 courtesy of the Naval History and Heritage Command).

No. 3 *Entrance to the Gosport Navy Yard*

Entrances to navy yards were often imposing structures and the Gosport Navy Yard's main gate upheld the tradition. The buildings on each side of the gate were used primarily as store houses and ran parallel to Portsmouth's Lincoln Street. The gate opened onto Water Street. These buildings were destroyed when the yard was burned on 21 April 1861, and were later rebuilt in a close approximation of their original appearance (Detail from Map No. 3).

No. 4 *The Steam Frigate Merrimack at the Norfolk Navy Yard*

Another view from *Harper's Weekly*, this image provides a realistic impression of *Merrimack*'s condition while laid up at Gosport. Note the house-like temporary structure covering her main deck, all to keep her interior dry while she languished on the Elizabeth. Her appearance conforms with McCauley's Senate committee testimony that "she was a dismasted hulk, with only her lower masts and bowsprit in, and without a particle of rigging to support her masts" (Photo # NH 59566 courtesy of the Naval History and Heritage Command).

No. 5 *The State Capitol, Richmond, Va.*

Thomas Jefferson is credited with the overall design of Virginia's State Capitol building, his inspiration arising from an extant Roman temple in Nîmes, France. It was in this building that Virginia's State Convention first assembled on 13 February 1861. While most of their subsequent deliberations occurred in the nearby Hall of the Mechanics' Institute, they were again convened in the Capitol building when on 17 April the delegates passed Virginia's ordinance of secession (Digital Id http://hdl.loc.gov/loc.pnp/ppmsca.34907 courtesy of the Library of Congress).

No. 6 *Governor Henry A. Wise of Virginia*

Virginia's former governor Henry Wise was one of the most outspoken delegates attending the State Convention. Although not an outright secessionist, Wise's philosophy of "armed neutrality" ensured his votes were aligned with their block. Meanwhile, on the sidelines, he organized his own personal insurrection to ensure Virginia would gain from the federal bounty at Harpers Ferry and Gosport Navy Yard. The role Henry Wise played in Virginia's capture of Gosport has largely escaped historians' attention (Digital Id https://hdl.loc.gov/loc.pnp/cph.3a21291 courtesy of the Library of Congress).

No. 7 *Secretary of the Navy Gideon Welles*

Welles' outlandish appearance was a product of his unkempt beard and mismatching wig. His three-year prior stint as chief of the Bureau of Provisions and Clothing gave him valuable experience with the Navy's ways. Fifty-eight years old when he assumed the role of Lincoln's secretary of the navy, Welles had been on the job for only forty-seven days when the Gosport Navy Yard was burned. Despite his rocky start, Gideon Welles was remembered as one of the best navy secretaries of the nineteenth century (Digital Id https://hdl.loc.gov/loc.pnp/cph.3b20114 courtesy of the Library of Congress).

No. 8 *Flag Officer Garrett Pendergrast*

A native of Kentucky, Pendergrast was a relatively senior captain serving as flag officer of the Home Squadron when the fates merged him with the destruction of the Gosport Navy Yard. He was married to Susan Virginia Barron, a proud Virginian, who may have influenced his behavior prior to the yard's loss. In a touch of irony, Pendergrast was the steam frigate *Merrimack*'s first captain; the records suggest he did not run a happy ship during his time in command (Image 529702 courtesy of the National Archives and Records Administration).

No. 9 *U.S. Sloop Cumberland*

Originally built as one of the "Gradual Increase" frigates, *Cumberland* was later razeed into a spar-deck sloop-of-war by cutting down her bulwarks and respacing her main deck gunports. By 1861 she was armed with twenty-two 9-inch Dahlgren smooth bores and a forecastle-mounted 10-inch pivot gun. In her recast configuration, *Cumberland* was a powerful man-of-war although she was hobbled by one glaring weakness—no steam power (Catalog # NH 61867 courtesy of the Naval History and Heritage Command).

No. 10 *Engineer-in-Chief Benjamin Franklin Isherwood*

Only thirty-eight years old when appointed the Navy's engineer-in-chief, Isherwood had already chalked up significant achievements as a prolific author of articles on steam engineering. Contemporary accounts suggest it was Isherwood who kindled Welles' focus on saving *Merrimack*. A tireless promoter for greater recognition of the naval engineers' role in the new steam navy, Isherwood was widely admired although his aggressive tactics earned him the dislike of a hard core of traditionalists (Digital Id https://hdl.loc.gov/loc.pnp/cwpbh.03238 courtesy of the Library of Congress).

No. 11 *Captain James Alden*

A direct descendant of a *Mayflower* pilgrim, the forty-one-year-old Commander Alden was put in the hot seat when Secretary Welles unexpectedly assigned him the duty of retrieving the steam frigate *Merrimack* from Gosport. Although he had seen much prior active service with the Coast Survey, he was unable to successfully organize the escape of one of the Navy's largest warships. Welles' final verdict was harsh. "At Norfolk all his heroic drawing-room resolution and good intentions failed him." (Photo # NH 52117 courtesy of the Naval History and Heritage Command).

No. 12 *Lieutenant Commander Thomas O. Selfridge, Jr.*

Young Selfridge was a newly minted junior lieutenant on *Cumberland* when she was ordered to guard the Gosport Navy Yard. Aggressive and bold, he played an outsize role in the days leading up to the yard's loss, later writing in his memoirs "I thought at the time … that the evacuation and partial destruction which followed was an irretrievable blunder." Eleven months later, Selfridge heroically defended his ship against the ironclad *Virginia*'s onslaught (Catalog # NH 2858 courtesy of the Naval History and Heritage Command).

No. 13 *Sectional View of the U.S. Steam Frigate Merrimac*

Merrimack was huge, one of the very biggest wooden war steamers at 256'-10" in length. This inboard profile emphasizes the ship's bulk and shows how her machinery was tucked deep inside her hull. Navigating a frigate of this size down the obstructed the Elizabeth River with only a skeleton crew, no pilot and propelled by Isherwood's "hastily patched up" engines would have been a risky gamble (RG 19, Bureau of Steam Engineering, Alphabetical File—Merrimack, Van Nostrand folder, Sectional View # 22 [Inboard Profile] courtesy of the National Archives and Records Administration).

No. 14 *Rear Admiral Hiram Paulding*

Paulding's exemplary service as a midshipman at the Battle of Lake Champlain was later counter-balanced by his lack of judgment when in command of the Home Squadron. His extra-legal removal of the filibuster General Walker from Nicaragua sparked presidential censure. Secretary Welles overlooked Paulding's slipshod performance during his first flying visit to Gosport and sent him back in an attempt to save the yard. But soon after arriving, Commodore Paulding hastily concluded that the Gosport Navy Yard must be burned (Photo # NH 47433 courtesy of the Naval History and Heritage Command).

No. 15 *U.S. Steamer Pawnee*

The steam sloop *Pawnee* was a one-of-a-kind. Designed with an exceptionally large breadth to permit her to carry a heavy armament on the light draft of ten feet, she was the best suited of any of the U.S. Navy's warships to undertake Paulding's relief expedition. Her wide, open main deck made it feasible to pack Colonel Wardrop's 350 Massachusetts Volunteers on board for the two-hour run to Norfolk (Photo # NH 45362-A courtesy of the Naval History and Heritage Command).

No. 16 *U.S. Steam Tug Yankee*

This humble New York-registered steam tug might have been a game changer at Norfolk if only Pendergrast or Paulding had grasped her potential. This photo shows *Yankee* later during the Civil War; as a commissioned naval gunboat she was credited with eight prizes. Admiral David Porter remarked of the debacle at Gosport, "One tug with a twenty-four-pounder howitzer on board, properly handled, would have been master of the situation" (Catalog # NH 25 courtesy of the Naval History and Heritage Command).

No. 17 *The Virginia (Merrimac) in Dry-Dock*

This image shows Gosport Navy Yard's graving dock hosting the Confederate States Navy ironclad ram *Virginia* during the final stages of her armoring. Commodore Paulding's failure to blow up this dock and to fully destroy *Merrimack*'s hull led to unexpected consequences eleven months later during the Battle of Hampton

Roads. If his team had achieved either one of these missions, the history of naval warfare would have been written differently (Catalog # NH 314 courtesy of the Naval History and Heritage Command).

No. 18 *Destruction of the United States Navy-Yard at Norfolk, Virginia*

This line engraving was published in *Harper's Weekly* in 1861 and is a relatively accurate depiction of the yard's destruction. The upper frame is a view from inside the yard looking toward the wharf and is dominated by the bulk of the two flaming ship houses. The sheers shown poking above the house on the right had been collapsed by this time. The bottom frame illustrates the escape of *Cumberland*, towed by *Pawnee* and assisted by the tug *Yankee*. The position of the receiving ship *Pennsylvania* to the right is artistic license; she would have been in the foreground (Photo # NH 59179 courtesy of the Naval History and Heritage Command).

***

No. 1 *Commodore Charles Stewart McCauley*

No. 2 *The Navy-Yard at Norfolk, Virginia*

No. 3 *Entrance to the Gosport Navy Yard*

No. 4 *The Steam Frigate Merrimack at the Norfolk Navy Yard*

No. 5 *The State Capitol, Richmond, Va.*

No. 6 *Governor Henry A. Wise of Virginia*

No. 7 *Secretary of the Navy Gideon Welles*

No. 8 *Flag Officer Garrett Pendergrast*

No. 9 *U.S. Sloop Cumberland*

No. 10 *Engineer-in-Chief Benjamin Franklin Isherwood*

No. 11 *Captain James Alden*

No. 12 *Lieutenant Commander Thomas O. Selfridge, Jr.*

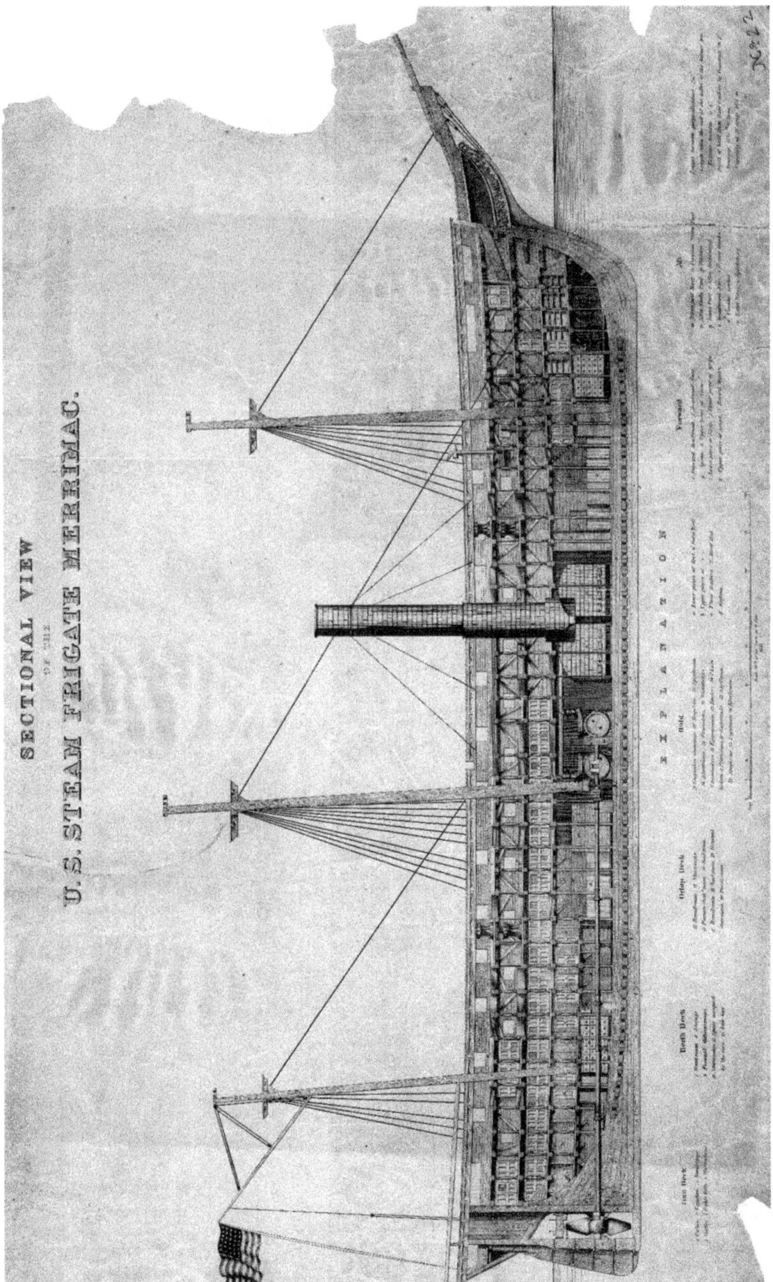

No. 13 *Sectional View of the U.S. Steam Frigate Merrimac*

No. 14 *Rear Admiral Hiram Paulding*

No. 15 *U.S. Steamer Pawnee*

No. 16 U.S. Steam Tug *Yankee*

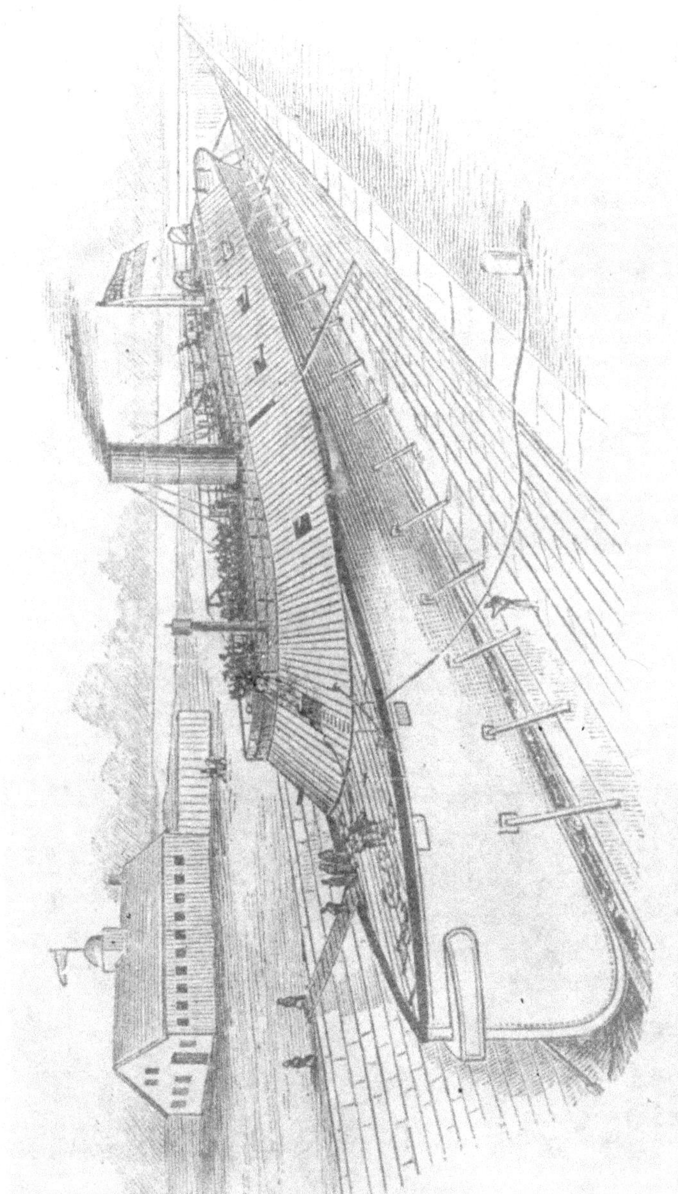

No. 17 *The Virginia (Merrimac) in Dry-Dock*

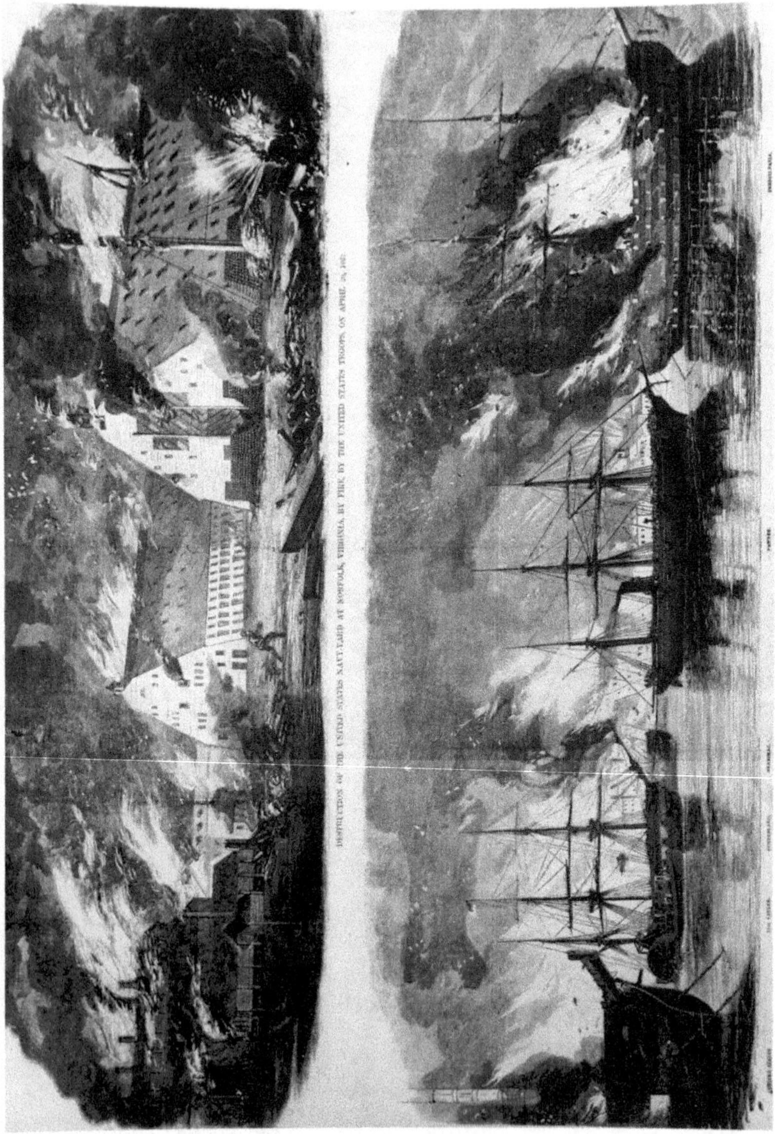

No. 18 *Destruction of the United States Navy-Yard at Norfolk, Virginia*

CHAPTER 11

Paulding's Flying Visit

On the morning of Wednesday, 17 April, the *Cumberland*'s Captain John Marston took up his pen to draft a report to his superior, Flag Officer Pendergrast, about a pair of suspicious sightings recorded the night before. Because of the foul weather and limited visibility, it was difficult to be certain of just what happened but this much was known. A tug had been spotted towing two light-boats down the river and then, thirty minutes later, a second tug was seen doing the same thing. Captain Marston sent two of *Cumberland*'s boats to investigate but both returned empty-handed; neither could say what had been done with the light-boats. But, as he wrote to Pendergrast, "previous reports led me to apprehend that they were to be sunk for the purpose of blocking the channel." Marston would have a lot to discuss when Pendergrast returned to his flagship.[1]

The second of *Cumberland*'s boats had been armed and placed in command of an aggressive young lieutenant, Thomas O. Selfridge, Jr. Following his return, Selfridge presented his captain with a bold proposal. He volunteered to take the brig-of-war *Dolphin*, then lying unrigged at the yard and without officers and crew, and man her with some of *Cumberland*'s hands, tow her down to Craney Island and use her to foil any future attempts to sink obstructions in the channel. Sometime earlier Commodore McCauley had taken it upon himself to arm the brig, placing four 32-pounder guns on board, and Selfridge's active mind saw her potential to stop the secessionist schemes. His captain and Flag Officer Pendergrast thought his scheme an excellent plan and said they would run it by the commodore.[2]

* * *

But further consideration of Selfridge's proposal was overshadowed by the arrival of Commodore Paulding. Just what time Paulding appeared that day or if any advance notice was telegraphed is a grey area. No record of a telegram can be found in the *Official Records* and Paulding confusingly testified to the Senate committee that he both arrived at and left Norfolk on the afternoon of 17 April.[3] Given that his testimony was read back to him and pronounced correct, it appears that Paulding spent only a few hours at the navy yard, a truly flying visit.

Commodore Paulding's first order of business would have been to introduce himself to Commodore McCauley and hand over the packet of Welles' orders written just the day before (for their full text, see Appendix 3). After describing the purpose of his visit, he no doubt offered some brief words explaining the views and purposes of the Navy Department. As he talked, one can imagine Commodore Charles McCauley reading his orders, digesting the latest "views" from Washington. He had now been instructed to remove no fewer than *four* major warships from the Gosport Navy Yard: a big steam frigate, two sloops and a brig. And before they sailed, not only did the Department expect them to be fully armed, but also to be loaded with stores and ordnance from his yard. But he had been given no resources to move the ships other than by stripping men from his own command and the only active man-of-war that could protect him, Pendergrast's flagship *Cumberland.*

McCauley quickly ran through the numbers in his head. Between the yard and its receiving ship, he had at most 180 seamen, and *Cumberland* was crewed by 350 officers, seamen and marines.[4] Skeleton crews for the four vessels Welles had ordered moved needed a minimum of 140 sailors and twelve officers.[5] McCauley was also aware Paulding had hand-carried copies of Welles' orders detaching four of Gosport's officers, but their reliefs were nowhere in sight.[6] So while he might be able to scrape together enough seamen to crew the four ships—thereby crippling the defensive efficiency of both yard and Pendergrast's flagship—there was only the recently arrived Commander Alden and perhaps three lieutenants to spare from *Cumberland* and the yard to officer them. Welles had presented him and Pendergrast with a Hobson's choice equally distasteful to both.

McCauley let his thoughts wander further. If the removal of the most valuable vessel, *Merrimack,* was conditional on "imminent danger pending," did that same logic not apply to the other three? And what more could the Department tell him about the "mob violence, organized effort, or any assumed authority" he was ordered to defend against at any hazard? McCauley was aware that his Virginian officers were keenly following the State Convention's proceedings. What would be his situation if Virginia seceded and claimed sovereign authority over the Gosport Navy Yard? Was the state of Virginia an "assumed authority" that warranted repelling with deadly force?

There was one more thing that rankled Commodore McCauley. The Secretary of the Navy had sent a courier to him carrying explicit orders backed by plenary powers. He well understood the need to transmit orders in this way during these troubled times, but could Secretary Welles—who he knew personally—not have sent a more senior officer than Paulding? Perhaps George Storer or Francis Gregory, both of whom were waiting for orders.[7] He had been slighted before by the Department when it ignored his request relayed by Captain Powell for reinforcements and, on another occasion, when it revoked his acting appointment of a boatswain. And now Paulding had arrived with these impossible orders.[8]

* * *

The three commodores, McCauley, Pendergrast and Paulding, put aside their private thoughts and got down to business. On his way into Gosport, Paulding couldn't help but noticing that "a threatening and hostile spirit seemed to pervade the vicinity of the yard. The public property seemed in some jeopardy." The decision was quickly made to move *Cumberland* from the naval anchorage to a position further up the river, opposite the first ship house so as to command "the entrance to the front gate and the western part of the yard." Some accounts put her final position opposite Ship House B, the second ship house, but most agree she was placed to cover the front gate. Once the move was complete, the sloop was moored "head and stern," or on springs, so she could turn her broadside to face a threat from any direction.[9]

As for who's idea it was to relocate *Cumberland*, Paulding testified during the Senate committee hearings that "it was determined by us" to remove the ship, then later said that McCauley had proposed it, which aligned with McCauley's testimony. But in Paulding's biography, he claims she was moved "at my suggestion." And to further muddy the waters Lieutenant Selfridge, who was an active participant in the outcome, remembered that the ship's move was made "on the advice of Commodore McCauley's pro-Southern officers."[10]

The removal of *Cumberland* to a position opposite the yard's front gate added a robust layer of protection to the Gosport Navy Yard but did so at the cost of surrendering federal control of the Elizabeth River. With the threat of her guns removed from the river's main passage to Norfolk, the secessionists' ability to further block the ship channel was immeasurably enhanced. Now there was nothing to stop them. True, *Cumberland* on springs made for a powerful gate-keeper, but that role could have been nearly as effectively performed by the 120-gun *Pennsylvania*, mud-bound across the water from the yard's entrance. From this analysis, Selfridge's attribution of responsibility for suggesting the move makes sense—the yard's Southern officers urged her relocation to further their stealthy campaign to isolate and intimidate McCauley. And the commodores played right into their hands.

Close on the heels of the decision to move *Cumberland*, the commodores determined to "send all the small arms that could be sent on board," a task that would be greatly simplified by the sloop's repositioning closer by the yard. Then Selfridge's proposal to guard the channel at Craney Island with *Dolphin* was aired. It was "cordially approved" by Paulding, McCauley grudgingly "gave his consent" and, as he had with other key decisions, Pendergrast remained a cipher. In combination, removing *Cumberland* to the yard and replacing her presence in the river with *Dolphin* made good military sense, but only if both elements were implemented. As will be seen, *Dolphin* never left the navy yard.[11]

Merrimack's situation merited the commodores' attention although their recorded deliberations seem to lack even the conditional decisiveness of their other evaluations—and she was the biggest prize of all. Paulding had noticed that a "feeble effort" was being made to coal her and that the big frigate's head was still upstream. He suggested her coaling be sped up, that she be winded, her battery placed on board and then be sent away but was met with the now familiar excuse that "such a movement would give offense to the people outside and that no more expedition in coaling could be made."[12] By this time Commodore McCauley's will to resist his Southern officers' urgings had been largely eroded and he was all too easily persuaded to avoid taking decisions that could excite the restive locals. And strangely, the commodores' later accounts and testimony about their fateful meeting on the afternoon of 17 April made no mention of either Engineer-in-Chief Isherwood or Commander Alden.

One can imagine Commodore Paulding glancing at the clock, his mind wandering to the Bay Line's steamer schedules and his return to Washington. But there was one more hurdle to clear, a key element of his mission: to satisfy himself on the reliability and fidelity of the yard's officers and men. While still in McCauley's office, Paulding was approached by Commander Richard Page, in charge of Gosport's now idled rendezvous, who sought him out on behalf of the Southern officers stationed at the navy yard. Paulding's testimony recorded his conversation with Page:

> He said that the officers were painfully situated; that their families and all their interests were there; that there was a very great excitement, and apprehensions were entertained of violence from the people outside the yard; that although it was known to him that

> the convention in Richmond was in secret session, and that Virginia would go out of the Union, he pledged himself, and was authorized by the other officers to pledge them, to stand by the commandant and defend public property with their lives. But he begged me at the same time to say to the Secretary of the Navy that it was very desirable to them to be relieved.[13]

Page's plea on behalf of his fellow officers must have struck an emotional chord deep inside the commodore; he later wrote that these were "men that I had known and esteemed for their honor from boyhood."[14] These same officers, who all bore commissions in the U.S. Navy, had pledged to defend the navy yard to their last, but they also admitted "it was very desirable to *them*" that they be relieved of their duty. Most remarkable of all, Page announced that Virginia would secede. As will be seen, his intelligence was flawless—on that same day, the Convention voted to approve an ordinance of secession. Of all of Gosport's Southern officers, Commander Richard Page was one of the best placed to know. He was a son of Virginia, lived in Norfolk, was married to a local woman and had relatives in high places.

Page's revelation hastened Paulding's rush to the door. "With this my mission ended and it was my duty to obey my orders and return to Washington. I lost no time in returning." Commodore Paulding left Norfolk that afternoon on the Baltimore boat.[15]

* * *

What had been accomplished during Paulding's brief visit? Pendergrast's flagship had been moved from the naval anchorage and moored opposite the yard's entrance, thereby leaving the secessionists unhindered in their campaign to obstruct the navy yard's only access to the sea. The transfer of small arms to *Cumberland* had begun. But soon after Paulding's departure, Commodore McCauley was prevailed upon by his officers to withdraw permission for Lieutenant Selfridge to take *Dolphin* and guard the ship channel. The excuse was that the young lieutenant's rashness "might precipitate a crisis which might throw Virginia to the Confederacy." Again, the commodore gave in even though it had been openly admitted that Virginia was leaving the Union.

What did the commodores achieve toward fulfilling Secretary Welles' orders? He had issued instructions for *Merrimack* to be put "in condition to be moved" and to get *Plymouth* and *Dolphin* "beyond danger of immediate assault at once."[16] And Page's warning should have spurred a vigorous discussion over measures to deal with Virginia's secession. Yet the commodores had done almost nothing to deal with any of these pressing matters.

During the later Senate committee hearings, Paulding was pointedly asked if McCauley raised "any complaints ... of an insufficiency of men ... there at the time" and he answered "I think Commodore McCauley might have expressed apprehension of an insufficiency of force." When next queried about whether McCauley stated "that there was any insufficiency of men to take the *Merrimack* out, or express any fears of a hostile attack on her after she should be taken out," Paulding replied "I think not, sir."[17] Could one then conclude that, yes, McCauley was worried because he didn't have enough crews to fulfill his orders, but he had no concerns about *Merrimack*'s sailing? The last inference flies in the face of later events.

The question of rank and seniority *did* come up during the Senate hearings and the officers' responses yield clues as to why the debacle latter occurred at Norfolk. When asked if he made any suggestions or gave any orders to Commodore McCauley to adopt for resisting an attack, Paulding replied, "I did not. I had no power to command him, inasmuch as he was my superior officer." And this answer came from Welles' hand-picked emissary supposedly vested with plenary powers. Paulding further clarified that "Commodore McCauley was the superior officer. He was Commodore Pendergrast's superior, and was in specific command of the yard." This begs the question of why Welles sent a junior officer to express the Department's views to the more senior commandant of the Gosport Navy Yard. Did he not realize that rank would stifle any open discussion of the situation and responsibilities? To complete the picture, when the ever-silent Commodore Pendergrast was later queried about what instructions he was operating under during the critical days of 17 through 20 April, he cited Welles' orders to McCauley dated 16 April. When asked "Did you receive a letter from the Secretary of the Navy of the same date, directed to you upon the same subject?" Pendergrast answered, "I think I did not."[18] If his testimony is to be believed, Paulding failed to deliver Welles' orders to the flag officer of the Home Squadron. The incoherence displayed by the three commodores regarding their orders, their priorities and their responsibilities is simply astonishing.

The Senate committee also clearly recognized the commodores' cumulative faults and rendered the following harsh judgment, drawing particular attention to Paulding's flying visit:

> Your committee cannot refrain from expressing the opinion that there was a total failure on the part of those three high officers to appreciate the circumstances by which they were surrounded, and

> the danger to the property they were instructed to defend, and to exercise that ordinary and prudent foresight which the government had a right to expect of officers of their age, character, rank, and experience. Nor can the committee understand why Captain Paulding, the special and confidential agent of the department, should have felt at liberty to leave, within a few hours after his arrival … without having first determined, in consultation with Captains McCauley and Pendergrast, upon some definite course to be pursued in the event of imminent danger to or an attack upon the yard.[19]

* * *

Meanwhile, back in Washington, Secretary Welles was showing signs of increasing anxiety about the situation in Norfolk. Although his orders carried by Paulding of 16 April contained instructions to place *Plymouth* "beyond the danger of immediate assault," on the following day he further specified she should be sent to Annapolis "at the earliest practicable period." But then Welles revealed he had yet to fully grasp McCauley's biggest problem—lack of crews to man all the ships he had been ordered to move—by asking if 130 officers and men could be spared for *Plymouth.*[20]

As if tapping into Welles' and McCauley's communications, that same day Commodore Breese of the New York Navy Yard reported to Welles that the crews he ordered sent to Norfolk on 11 April—nearly a week before—would *probably* be ready for transfer in about three or four days. Reflecting the uncertainty sweeping the nation, Breese requested that the Department revert back to him if there were any "changes in its intentions."[21]

And doubling down on his sense that Norfolk might need help sooner rather than later, Secretary Welles ordered Lieutenant Ronckendorff, in command of the side-wheel sloop *Water Witch* at the Philadelphia Navy Yard, to "sail immediately for Norfolk." Because of the recent mobilizations to Charleston and Pensacola, she was the only war steamer available to a navy stretched to its limit. But Welles was too late; *Water Witch* had left for Fort Pickens before his order could reach her.[22]

* * *

Wednesday 17 April had been a very full day for Commodore McCauley. First the news about the obstructions in the river, then came the rush of Paulding's visit and all the meetings, decisions and orders flowing from a raft of decisions, one of which was moving *Cumberland* up to guard the yard's front gate. On the other hand, he had been encouraged that the yard's new naval constructor, John L. Porter, had just reported for duty. A native of Portsmouth, Porter would no doubt prove useful in the days ahead. Little could McCauley, or anyone else for that matter, dream of how Porter, Gosport and *Merrimack* would become eternally entwined.[23]

And now when McCauley thought things couldn't get any busier, Engineer-in-Chief Isherwood and Chief Engineer Danby called on him. The two engineers reported that *Merrimack*'s repairs had been completed and her machinery was ready for steam. Isherwood then said, "Should I fire up at once?"

The stress of having to deal with the rapidly unfolding and menacing series of events that had descended on his navy yard was beginning to tell on the old commodore. Since last Sunday the engineers had been hard at work, day and night, and McCauley had yet to set foot on the steam frigate. Perhaps he should inspect her. But from his horizon, the "immediate danger" threshold that Secretary Welles cited as the trigger to force her departure from Gosport had not been reached. Besides, her battery was not on board and he had yet to hear from Commander Alden.

"No," answered McCauley, "not this afternoon." He paused and looked away lost in his thoughts. "If you have steam up tomorrow morning, that would be time enough."

Isherwood and Danby had received their answer. It may not have been the one they expected, but for the moment it would do. The two men walked back to *Merrimack* and calling together their engineers and firemen, ordered that they begin regular engine room watches. Instructions were issued to light off the furnaces at daybreak.[24]

* * *

CHAPTER 12

Virginia Votes for Secession

On the morning of Wednesday, 17 April—a fateful day for the Old Dominion—Henry Wise hovered in the lobby of Richmond's Exchange hotel. His ears and eyes were keenly tuned to detect the clerk's confirmation that his expected telegram had arrived. A couple of blocks away the State Convention had begun its second day in secret session. Delegates from the western and trans-Allegheny regions were that morning's only speakers, and they could be reliably expected to urge further delays in the name of Virginia's unity to proceedings that had now carried on for over two months.[1]

"Governor Wise!" shouted the clerk, waving a piece of paper. Henry Wise grabbed the telegram and saw it was from Captain Imboden. It had been sent from Gordonsville, and said he was there with his volunteers under General Harper, and confirmed they were pressing on to Harpers Ferry. Excellent news! His spirits now buoyed, Wise hastened across to the Capitol building and the Convention within.[2]

The Virginians who conspired the night before with Henry Wise to instigate their private rebellion had agreed that all of their dispatches would indicate the volunteer militia's destination as the "Portsmouth Navy Yard" rather than its true target, Harpers Ferry. This cover was meant to act as a diversion in the event their communications were compromised or leaked. Telegraph offices were public places and it was all too easy for an unguarded remark to be picked up by those of a different persuasion.[3]

By the time Wise had reached his seat at the Convention, a substitute ordinance proposed by Robert Scott had been narrowly defeated by 79 to 64. That set the stage for a vote on William Ballard Preston's ordinance of secession. It was in this sea of expectancy that Henry Wise, long a master of dramatic confrontation, claimed the speaker's podium. For some weeks before, breathless delegates would rise up brandishing telegrams that they insisted heralded a dramatic new development but were later derided as

"Another Democratic alarm!" [4] Now Henry Wise brought them something very different. Drawing a large horse pistol from his belt, he lay it before him. Next, he pulled a watch from his pocket. His eyes afire and shaking his watch, Wise proclaimed that events were transpiring this very moment that would try any man's soul. Waving Imboden's telegram, he addressed the Convention:

> Mr. President, I arise to announce no "Democratic telegram," but to say to you and this body that I know that armed forces are now moving upon Harpers Ferry to capture the arms there in the arsenal for the public defense, and there will be a fight or a foot-race between volunteers of Virginia and Federal troops before the sun sets this day![5]

The former governor then claimed that federal property and the navy yard in Norfolk were being seized by his insurgent forces. His gaze intensely sweeping the sea of astonished faces before him, Wise called upon the Convention to join him in support of the volunteers and sanction their forward movement.

The delegates arose in an uproar. Wise's biographer described the scene: "If a hand grenade had been thrown in among the members, it could not have caused more consternation."[6]

In the excitement that followed, Wise engineered a summons to the governor's chamber and was soon called away from the convention floor. Now face to face with Governor Letcher, Henry Wise seized the moment to redeem a promise the governor had made the night before, and asked that he sanction the volunteers' movement. Presented with a *fait accompli* and fully grasping the push toward secession was inexorable, Governor Letcher agreed and issued the necessary orders.

Returning to his seat at the Convention, Wise quietly listened as Unionist delegates cried out in protest at his usurpation of law and order, and questioned what appeared to be a treasonous motivation for ordering Virginia's volunteers to march on Harpers Ferry. Who had devised such a reckless scheme?

Henry Wise stood up and, without hesitation, claimed that he alone had organized the movement. He then threw out a challenge. Now that their fellow Virginians were marching to seize arms for their own state's defense, would the Convention not give them its support? Next followed a famous exchange when the delegate from Staunton, John Baldwin, refused to sanction such a rash act that hadn't first been approved by the people, taunting that those who had done so "made for themselves a bloody bed" and should "lie on it, and take all the consequences." Wise swiftly rose and, in a torrent of oratory, informed Baldwin that those he consigned to such a sanguinary fate were, in fact, his Augusta county constituents, men like General Kenton Harper and Captain John Imboden. "Shall they be doomed, unsupported, to bloody beds?" At that moment of extreme emotion, with Baldwin on the ropes, a delegate burst forward down the aisle, pushing chairs aside. Reaching for Wise's hand, he cried out "I don't agree with you, I don't approve your acts; but I love you! I love you!"[7]

That Henry Wise powerfully affected men is beyond doubt. He had inspired a hastily gathered cabal of conspirators to mount an armed assault on a United States arsenal. He had approved Norfolk's vigilance committee's obstruction of the channel to the Gosport Navy Yard. Aided by his horse pistol he had jolted Virginia's Convention into the lap of the secessionists. And he had strong-armed Virginia's governor into officially sanctioning his personal insurrection. By doing whatever it took, by exerting his reservoir of authority over his fellow Virginians, Henry Wise had dragooned his state into embracing his long-advocated stance of "armed neutrality."

But Wise's work was not yet done. Virginia was still in the Union and for that to change, the Convention had to pass the ordinance of secession. But before the delegates could cast their votes more debate was needed, so Henry Wise again swung into action. His focus on that final afternoon of the Convention's closed session was a moderate delegate from western Virginia, George Summers. Wise's stinging rebuke of Summers was vividly described by an eye witness:

> Before he arose, I noticed his suppressed agitation. Ex-President Tyler, who knew what was coming, turned his chair about ten feet in front of Wise, with his back to the president of the Convention. As Wise proceeded with his arraignment of Summers, Mr. Tyler lost control of his feelings and tears trickled freely down his cheeks. The speaker was supernaturally excited. His features were as sharp and rigid as bronze. His hair stood off from his head, as if charged with electricity. Summers sat on the left of the chair, white and pale as the wall near him. It was the most powerful display of the sort I ever witnessed.[8]

But finally, the talk ended. The Convention had first been assembled on 13 February 1861, over two months before. Countless votes had been cast and late in the afternoon of 17 April—4:15 p.m. to be exact—the President of the Convention, Mr. Robert Montague, announced the result of yet one more vote, this one on William Ballard Preston's ordinance of secession. It passed by 88-55.[9]

Virginia had *not* seceded from the Union, but it *had* passed an ordinance that set in train the state's secession pending its endorsement by a popular vote to be held in little over a month. However, from the enthusiastic reception that greeted the Convention's vote, it was hard to escape the conclusion that Virginia was already out of the Union.

* * *

Although the fateful vote had been cast, the Convention still carried on through the afternoon and, indeed, for months later until the Old Dominion's business of transforming itself into something other than an integral element of the United States was finished. Late on that day, a lively debate occurred between the ever-ready Henry Wise and the by now recovered John Baldwin over the people's power allowed by military necessity versus the need to follow constitutional form. Extracts of Wise's addresses are reproduced to unequivocally demonstrate that the Gosport Navy Yard was targeted by the Convention's secessionists. His second address closely resembles a report given the day before by George Wythe Randolph:

> Is there any man here, when the car of war is rushing over the people themselves … when the edge of every bay and estuary in the bay is endangered by invasion; whilst your steamers are seized in New York; whilst your people have risen up and sunk hulks in the channel at Norfolk—will you tell me, sir, whether the *salus populi* does not require us to take the responsibility of doing *ad interim* whatever we can do for their defense …[10]

> Let the people say, we have seceded; our Convention has advised us to secede. In order to defend ourselves, in case we elect to secede, we will take the arms necessary to defend ourselves—we will take the forts that now threaten our lives and liberties; we will take the Navy-yard that holds all the ship timber, the best in the United States. If we take the Navy-yard now, we will have plenty of arms and ammunition. If we do not take it, when we vote to secede, we will have no ships, no muskets, no ordnance, no powder. Yes, sir, there lies 3,000 barrels of powder, and the gentleman's argument is that you can do nothing in self-defense; but must permit the enemy to invade us without raising a hand in resistance.[11]

These passages clearly display that Henry Wise and his fellow conspirators possessed first-hand knowledge of the sinking of light-boats to obstruct the Elizabeth River, of exactly how many barrels of gun powder were held in the Navy's magazine at Fort Norfolk and of the plentiful small arms, ordnance and military supplies stored within the Gosport Navy Yard. Wise and his cohorts may not have been professional soldiers—at least not yet—but they had grasped one of the fundamental tenets of military science: they knew their enemy and knew him well.

* * *

Before closing out this account of the momentous proceedings that occurred in Richmond on 17 April, an examination of the language employed in the just-passed ordinance of secession is vital to gain a full understanding of how it affected Virginians and, in particular, Virginian officers of the U.S Navy. Two key elements of the collective "Ordinances" command attention, the first and ninth ordinances:

> No. 1—An ORDINANCE to repeal the ratification of the Constitution of the United States of America, by the State of Virginia, and to resume all the rights and powers granted under said Constitution.
> The people of Virginia ...
> *And they do further declare,* That said Constitution of the United States of America is no longer binding on any of the citizens of this state.
> This ordinance shall take effect and be an act of this day, when ratified by a majority of votes of the people of this State ...
>
> No. 9—An ORDINANCE to call the Volunteers into the Service of the State and for other purposes.
> *Be it ordained,* That the Governor of the Commonwealth ... shall immediately invite all efficient and worthy Virginians and residents of Virginia in the Army and Navy of the United States to retire therefrom, and to enter the service of Virginia, assigning them such rank as will not reverse the relative rank held by them in the United States Service, and will be at least equivalent thereto.[12]

First, the Ordinances' No. 1 Ordinance made perhaps the most consequential statement of the Ordinances beyond repealing Virginia's ties to the Union, that of dissolving the connection between the citizens of Virginia and the United States' Constitution. For a Virginian officer serving in the U.S. Navy and who elected to remain a citizen of Virginia, this declaration effectively released him from the oath that bound him to the

United States service. The corollary was that if a Virginian naval officer chose to follow the flag, he would need to renounce his Virginian citizenship or, at the very least, claim citizenship of a state that was still within the Union, if he were to remain "loyal" to the U.S. Navy.

Second, the No. 1 Ordinance stated that the provisions of the ordinance shall only take effect *and be an act of this day* when ratified by the people. So if a person were to take immediate action based on the promises held out by the Ordinances—say by resigning their commission in the U.S. service—they would be betting that the Ordinances would be endorsed by the scheduled election. For all practical purposes, this caveat was largely if not completely ignored in the euphoria following the Ordinances' passing. Jumping into the future, the Ordinances were ratified by the citizens of Virginia on 23 May.

Third, the Ordinances' No. 9 Ordinance offered, in the name of Virginia's governor, a "golden bridge" to Virginian U.S. Navy officers who chose to remain citizens of their state and who wished to continue their careers as naval officers. This offer must have been immensely reassuring to men who had since boyhood, invested their lives in a profession that didn't readily allow for changes of allegiance. The Convention delegates knew well what they were doing when they drafted No. 9 Ordinance, the only ordinance other than No. 1 to be passed on 17 April. It was a magnet to attract all serving U.S. officers.

The effects of the Ordinances reached far beyond the Old Dominion's secession. Virginian and other Southern officers were forced to make decisions their Northern peers never faced, to choose between their state or their flag. Events had rendered it impossible to claim loyalty to both. Each of them was confronted with a life-altering decision, a crisis of loyalties. Taking this a step further, one historian has characterized their dilemma as a "choice of treasons."[13] Which of their loyalties were they to renounce? A person's hierarchy of allegiances begins with the family, then one's community, then state and finally country. Was it treasonous to place loyalty to home and state above that to a nation, the bonds to which one's fellow citizens had dissolved by a popularly elected convention? Historians who carelessly apply the words treason and traitor should pause and place themselves in the shoes of those who faced making such a soul wrenching decision. Of the 147 Virginian naval officers who chose to "go South," only seventeen did so before Virginia passed its Ordinances. For the remaining 130 "reluctant rebels" who confronted the terrible crisis of loyalties, their homes and state won out against what was now viewed as a menacing federal Union.[14]

Could Gosport's Commander Richard Page possibly have known "that Virginia would go out of the Union" at the time he had his Wednesday afternoon conversation with Commodore Paulding? The answer is "no," but he probably had been reliably informed that the momentum for secession was becoming irresistible. Recall that earlier in the day an alternate ordinance was only narrowly defeated and that the passage of the final ordinance of secession was only known to the delegates late that afternoon.

The Old Dominion's vote to secede from the Union irrevocably impacted the lives of every Virginian and, in time, all who lived in the North and South. Its crisis of loyalties would soon slam the Gosport Navy Yard.

* * *

At nearly the precise moment Virginia's delegates passed their ordinance of secession, a sea-stained steam tug nudged her bow into the wooden bollards of a Norfolk pier. Her name was *Yankee* and her home port, New York, epitomized the maritime strength of the Union. How was it that this tug docked at Norfolk at 4 p.m. on the 17th of April?[15]

Yankee was one of the fleet hastily chartered by Gustavus Fox in support of his failed expedition to relieve Fort Sumter. The only one of his three tugs to actually reach the Charleston bar, *Yankee* arrived just hours after Fox had sailed for New York on *Baltic* with Major Anderson and his garrison. Her mission over before it began, *Yankee* turned her bow north and started the long steam back to New York. But the foul weather that had plagued her outward voyage continued to pound her, causing the loss of her smokestack. Short of coal and needing repairs *Yankee* sought refuge in Norfolk.[16]

Although she had been chartered as a mercantile steam tug, in preparation for her Fort Sumter service Fox fitted *Yankee* with a device to "throw hot water." Her voyage was conducted under the terms of a government charter, and her Captain Charles Germain was competent enough to be later warranted as an acting master. *Yankee* was a good-sized vessel, a side-wheeler measuring 146 ft in length and of 328-ton burden. In her later career, she was taken into naval service, appropriately enough first under the command of Acting Lieutenant Gustavus Fox, and would go on during the Civil War to seize or destroy eight prizes.[17]

For the moment, *Yankee* was just another tired steamer seeking safe harbor. But it was her ill luck to have arrived just after news of Virginia's secession reached Norfolk. Her name, home port and accounts of her involvement with Fox's Fort Sumter relief effort made her an obvious vigilance committee target. Fortunately, *Yankee*'s captain and crew kept up their guard, and on "finding a movement afoot to seize the boat, proceeded to the navy-yard and placed the steamer under the guns of the vessels there."[18] The next morning, Commodores McCauley and Pendergrast awakened to find a government-chartered steam tug tied up at the Gosport Navy Yard.

* * *

It wasn't until the evening of Wednesday, 17 April, that Secretary Welles read the telegraphic dispatches of the Washington *Evening Star* and learned the alarming but not unexpected news that Virginia had passed its ordinance of secession.[19] But his even greater surprise came the next day when Commodore Paulding unexpectedly returned to the Navy Department. In what must have been an anxious and probing exchange, Paulding quickly filled in Welles on the details of his visit to Norfolk. The secretary later recalled their conversation, beginning with the plight of the yard's Southern officers:

> Some of the younger officers, who belonged in Virginia or the South, had expressed a wish to be relieved from duty at the yard in anticipation of difficulty with the insurgents, among whom were their kinsmen and neighbors, with whom they preferred not to come in collision; but all were ... patriotic, deprecated hostility, and were governed by honorable motives. Commodore McCauley he indorsed as faithful, competent, and to be trusted. He was seconded by Commodore Pendergrast ... [Paulding] said he had consulted freely and fully with both those officers, had made some suggestions and assented to others made by them, and was so well satisfied that the workmen were reliable and that the public property was in good and trustworthy hands, that he thought it unnecessary he should remain, but that it was best he should return to Washington and make report in person.[20]

Although Welles found Paulding's report "more favorable than I expected," Welles also "greatly regretted that he did not remain and act for the Department, and so informed him." In his diary, Welles blamed himself for not issuing Paulding "explicit written orders to that effect."[21] His biggest shock was undoubtedly the blunt truth that nearly all the officers at the Navy's most valuable dockyard had claimed they were "painfully situated" and had requested reliefs. Whether they would follow Virginia out of the Union was the unanswered question.

Secretary Welles and the commodore lost no time in going to the president, who was then in a cabinet meeting. Paulding again recited his report, the main topic quickly turning to the message from Gosport's

Southern officers. Lincoln listened attentively and responded that he felt their request was reasonable and "that the officers must be relieved by others from the North." Secretary Welles was directed to issue the necessary instructions.[22]

Only three days before Welles had signed orders relieving from their duties Gosport Navy Yard's executive officer, Commander Robert Robb, its first lieutenant, Charles Spotswood, and its ordnance lieutenant, George T. Sinclair. All three were Virginians. Both Robb and Spotswood had been replaced by Northerners and Lieutenant John Irwin, a Pennsylvanian, had been sent to fill an opening on the yard's receiving ship. But they were just the tip of the iceberg. Also demanding relief were Commanders Richard Page, John Tucker, Thomas Rootes and Arthur Sinclair. And several of the yard's other lieutenants, its warrant officers—gunners in particular—also were candidates for replacement. It was a complicated process and it would take time, a precious commodity that was quickly running out.

Others in Washington had also gotten wind of Virginia's move toward disunion and reacted in accordance with what little information they had heard. Commander John Dahlgren, still stationed at the Washington Navy Yard, wrote in his diary for Thursday, 18 April:

> It has now leaked out that Virginia seceded on Tuesday, secretly, in order to grab the public property that is within her borders; and it is rumored that the Norfolk Yard and Harper's Ferry have been seized. Every one believes, too, that a body of men are on the way to take Washington, and the alarm is intense.[23]

* * *

If Secretary Welles couldn't quickly replace the Gosport Navy Yard's officers, he could bend every effort into tackling the impossible: finding Commodore McCauley the crews so desperately needed to move his ships to safety. He took action that same day. First, Welles tapped Baltimore's receiving ship, ordering its Commander Hunter to draft fifty recruits and send them that evening to Norfolk on the Bay Line's evening boat.[24] Second, he shot a telegram to prod Commodore Breese at the New York Navy Yard: "Send all the available recruits and ordinary men of the yard, to 250, to Norfolk immediately, by chartered boat if necessary."[25]

Beyond officers and men, Welles could strengthen the Navy's hand in Norfolk by sending a steamer to reinforce *Cumberland* and give Commodore Pendergrast the means to control the Elizabeth River. But at this point in time the peacetime Navy was overwhelmed—all the Home Squadron's steamers were fully committed. So the secretary had to think outside the box. His solution was to charter a merchant steamship. It may well have been Commodore Paulding who proposed the idea; it was he who wrote the order that day to Samuel Francis Du Pont, commandant of the Philadelphia Navy Yard, to do just that. Welles also dispatched Lieutenant Maxwell Woodhull to "proceed without an hour's delay to Philadelphia," where he reported to Du Pont who by then had chartered the fast steamer *Keystone State*. In a display of what a spirited commandant could accomplish, the next day Du Pont informed Welles that *Keystone State*—a 1,364 ton 219-foot long vessel—would be ready for sea that night "with her a guard of over fifty marines, a body of fifty seamen ..., four 12-pounder howitzers ... with ammunition."[26]

Secretary Welles deserves full credit for his vigorous and creative responses to the news of Virginia's secession and Commodore Paulding's troubling report. But if he thought he could rest on his laurels, he was sadly mistaken.

* * *

CHAPTER 13

The Commodore's Decision

At daybreak of 18 April, *Merrimack*'s engine room came to life for the first time in fourteen months. Thick coal smoke began wafting from her tall stack and within an hour, the river's muddy waters churned to her propeller's beat. Excellent—*Merrimack*'s engines were now turning under steam. It had only been four days since Engineer-in-Chief Isherwood had walked through the navy yard's gates. He now had the confidence to report to Commodore McCauley that the big frigate's machinery was ready for service.

Isherwood judiciously waited until Commodore McCauley had settled into his office. At 9 a.m. he called on the commodore, proudly reporting *Merrimack*'s engineering department was "ready for leaving." Chief Engineer Danby, his assistants, their firemen and coal heavers were all on board and the engines were working. All that was needed was the commodore's "order to cast loose and go." To Isherwood's surprise and dismay, McCauley informed the eager engineer that he hadn't yet made up his mind about sending the vessel. He would let Isherwood know his decision in a few hours. Isherwood called the commodore's attention to the fact that Secretary Welles' orders were peremptory. Then he raised the matter of the channel obstructions, giving his opinion that they could easily be passed, and any delay would doubtless see them increased. To that McCauley repeated his position, closing the interview by again saying he would decide during the course of the day.[1]

Isherwood's recounting of his first exchange that day with Commodore McCauley, which was used to construct the above narrative, displayed a level of aggression unusual for a staff officer. Perhaps the news of Virginia's secession had emboldened Isherwood. *Merrimack*'s acting captain, Commander Alden, was the one who could rightly be expected to adopt a confrontational tone when arguing an operational issue such as leaving port, not the engineer in charge of refurbishing her machinery. Both officers' orders were clear—they had been instructed to report to Commodore McCauley for duty as *he* may direct.

A not-so-easily dissuaded Engineer-in-Chief Isherwood quickly found Commander Alden and asked him to go on board *Merrimack*. The two officers descended into her engine room where Isherwood pointed to his assembled black gang—engineers, firemen and coal heavers—then waved at the engines' two enormous pistons panting steam, their cranks and rods flashing in the gloom. Leaning toward Alden, over the din he shouted, "The vessel is ready to go; my part is done!"[2] Alden nodded his acknowledgment, then climbed back to the main deck and its cooler air. Soon after Isherwood adopted a questionable tactic: he kept the ship's "engines running at the dock all day as a visible sign that the frigate was ready to go."[3] He may have sincerely believed his action would positively influence the commodore; it most certainly broadcast *Merrimack*'s pending escape to all Southern sympathizers inside and outside the yard.

A newly heartened Commander Alden strode off to see the commodore, who was then in the company of Commodore Pendergrast and *Cumberland*'s Captain Marston. Alden reported that *Merrimack* was ready and he only waited for McCauley's permission to take her. For the second time that morning, an anxious officer had pressed the commodore for a decision, and for the third time, he replied he hadn't yet made up his mind. McCauley paused, then offered his view that she might make a formidable defensive battery to protect the yard. He also thought the obstacles in the river would stop the big frigate. Alden countered that *Cumberland*'s Lieutenant Murray had sounded the channel early that morning and found an opening large enough for *Merrimack* to pass. Sensing he had gained the commodore's attention, Alden appealed his case in the strongest terms, citing Secretary Welles' orders undeniable intent for him to take charge of *Merrimack*. He turned to Pendergrast and Marston for support. A tense moment of quick words followed and then came the commodore's hesitant decision. "Take her, and if you don't get her out burn her."[4]

* * *

What should have been a hugely satisfying moment for Commander Alden quickly turned into a bed of ashes. The drag exerted by Gosport's Southern officers, throwing up every "red tapeism" they could muster, had throttled Alden's attempted departure long before he could cast off a single line. His five days of effort to prepare *Merrimack* for removal had been stymied at every turn. As of that morning, the 40-gun frigate had not a single cannon on board. Alden asked McCauley for two field pieces; yes, go see Commander Tucker. In a yard that hosted a total of 179 seamen and landsmen, Alden had yet to gather a crew. He was reduced to begging for a lieutenant and thirty men; yes, Captain Marston could help. The ship was still headed up-river and needed winding; yes, go see the yard's first lieutenant. And despite persistent inquiries, offers of cash and a navy job for life, neither Alden nor anyone else could persuade a local pilot to guide the big ship down the Elizabeth River, a restricted waterway now strewn with obstructions. Both Isherwood and Alden had brashly reported that all was ready for *Merrimack*'s departure, but the facts tell a different story.[5]

And things got worse during the hours that followed. The ability of the Gosport Navy Yard to make good *Merrimack*'s deficiencies and get her underway was rapidly collapsing. On that same Thursday, the 18th of April—not twenty-four hours after Virginia passed its ordinance of secession—nearly all of Gosport's senior Southern officers resigned from the U.S. Navy. As each hour passed, one after the other of Commodore McCauley's officers signed their letters of resignation, each man severing not only a lifetime's allegiance to the United States but also dissolving the executive structure so necessary for a navy yard to function (see Appendix 2 for a graphic chronology of Gosport Navy Yard's officer resignations, dismissals and detachments).[6]

Commander Alden vividly described the "thickening atmosphere of treason" around him, indicating he sensed many of the yard's officers were "going South." At that instant, however, he couldn't have known of the full extent of their mass exodus. But he did have his orders *and* the commodore's permission to take *Merrimack*, so despite the hurdles he pressed on. Alden later recalled the wall of obfuscation he ran into:

> I soon followed [Marston] to make my preparations for starting, and sent an order on board to be ready to cast off in two hours. … I then went in pursuit of the ordinance and executive officers of the yard. The first named threw all sorts of obstacles in the way of my getting the guns, and the other could not be found. The first lieutenant, in answer to my application to have the ship winded and fasts singled … said I must apply to Commander Robb, who I soon after met coming out of the commodore's office with the intelligence I could not have the ship.[7]

All of the yard's officers mentioned in Alden's account, Commanders Robb and Tucker (ordinance) and Lieutenant Spotswood (executive officer) had already resigned or would soon submit their resignation letters that day. Further, Robb, Spotswood and Tucker's ordnance lieutenant, George Sinclair, had only a few days before received orders detaching them for other duties. Since then they had all been biding their time, awaiting their replacements; now they were little more than walking ghosts. Secretary Welles' well-intentioned efforts to replace Gosport's Southern officers helped trigger a disruption of the yard's executive authority as it entered its moment of crisis. And then the winds of Virginia's secession blew through the navy yard, sealing *Merrimack*'s destiny.

As Alden just heard, his mission had failed. Unwilling to trust Robb's word, he "called upon the commodore and found it was true; the fatal order had been given."[8] Alden's account of this pivotal decision displayed no emotion in response to what must have been a crushing blow to his professional esteem. He was also silent about whether he just "struck away from the yard" or pushed back and demanded an explanation from the commodore. McCauley was not so reticent when he later recounted his rationale for refusing to allow *Merrimack*'s departure:

> I had no men to man and defend her in case of an attack, and as I entertained doubts of her being able to pass the obstructions placed in the river the night before, the existence of which I only learned that day, besides its being impossible to procure a pilot, I concluded that it would be injudicious to allow her to proceed. It would, in my opinion, have been incurring too great a risk. I therefore preferred exercising my own judgement in the matter, and gave the orders for her detention, as the responsibility rested wholly on myself.[9]

Seven months later Commander Alden, who would have had the most to gain had he succeeded that afternoon, tacitly endorsed McCauley's decision in a letter to Secretary Welles:

> Seeing me alone and single-handed, with no pilot … might he not be excused, under the circumstances, for doubting my power to carry that large ship safely through a narrow channel and across a barrier which these traitors had represented to him as impassable?[10]

Secretary Welles was far less charitable about Alden's behavior that day. Confiding in his diary, Welles expressed his unvarnished contempt:

> At Norfolk his heroic drawing-room resolution and good intentions failed him. He had not the audacity nor the moral courage to meet his professional brethren who had those qualities and were determined to sustain the Secessionist cause.[11]

If Commander Alden had timidly accepted his defeat, the same could not be said of Engineer-in-Chief Isherwood. At 2 p.m. that day, after waiting for the signal that *Merrimack*'s departure was imminent, Isherwood marched across the yard and confronted the commodore. McCauley confirmed his decision to retain the vessel and instructed Isherwood to "draw the fires" and stop the ship's engines. Amazed at what he had just heard, the feisty engineer-in-chief waved his peremptory orders and told McCauley in no uncertain terms that the big frigate must be dispatched at once. Furthermore, she should tow *Germantown* out behind her and could do so "with perfect safety." But the commodore held firm. And so Isherwood, realizing like Alden that his mission was ended, pulled out his orders and requested that the Commodore endorse them.[12] For Isherwood it must have been a savagely bitter moment: he *had* repaired *Merrimack*'s engines and did so faster than anyone believed possible. Now it was all for naught.

When later queried during the Senate committee hearings about Isherwood having "expressed the conviction that she could then be taken to safety," McCauley answered with a reply shaped by both their differences in ranks and the line versus staff officer divide, but also one that revealed the factors influencing his decision:

> He may have done so but I have no recollection of the fact. It was no part of his duty to do so. He was sent there charged with the special duty of repairing the engines, and when he had done that it was his business to report to me. I remember that he and Commander Alden were anxious to take her out; but I looked on all sides of the question, and considering the danger, and being every hour expecting reinforcements, I judged it best not to move her.[13]

* * *

Both Alden and Isherwood's accounts of McCauley's behavior as expressed in their reports to Secretary Welles were business-like, describing their exchanges with the fidelity expected of naval officers. Years after the event Isherwood wrote a "thorough and most carefully written account of the affair" that painted a darker, more troubled image of McCauley, shades of which were echoed in Secretary Welles' more recent diary entries for that period:[14]

> The Commodore was in a state of complete prostration. He sat in his office immovable, not knowing what to do. He was weak, vacillating, hesitating, and overwhelmed by the responsibilities of his position. He … could not be made to give any order or take any

> action. He looked vaguely at me, nodded his head, but said nothing. He behaved as though he were stupefied.[15]

Many of Isherwood's characterizations of McCauley ring true. Captain Du Pont had previously branded him as weak, when history and the Navy would have been better served if he had been strong. Yes, he did hesitate when he should have been more decisive, in the mold of Lieutenant (later Admiral) David Porter. And no one could disagree that he was cautious when the hour called for boldness. But vacillating is unfair. McCauley made a decision to hold *Merrimack* back and he stuck to it—more about that later—but it wasn't the decision that Alden or Isherwood wanted to hear. And all agreed, to use Isherwood's words, that the commodore "was brave ... and considered himself bound to struggle to the last."[16]

McCauley's state of prostration, sitting immovable and behaving as if stupefied may have been due to a cause hinted at by Welles, that he was "under the influence of liquor and bad men."[17] If he was drinking, he was in good company. Captain Wilkes later recalled of those who accompanied Paulding to rescue Norfolk that "half the officers had been taking their drams."[18] Did Navy officers drink on duty? Yes. Considering the pressure he was under, it would not be surprising if McCauley had imbibed a nip or two.

But the key phrase of Isherwood's description, that McCauley was "overwhelmed by the responsibilities of his position," has the most traction of all. Very few of the Navy's officers were prepared to endure the ordeal McCauley suffered through, a trial he bore under increasing isolation, carrying the total responsibility for making decisions that could—and eventually did—cost the government dearly and ruin his reputation. That his battle was entirely psychological and non-kinetic made it even more stressful for the old commodore. Accounts say McCauley had his "brave little boy at his side" during those difficult days, his thirteen-year-old youngest son Edward, but the lad was not somebody he could confide in.[19] And McCauley's relation with the one officer who should have been a source of support, Commodore Pendergrast, appears to have been distant and constrained by their differences in rank and command authority.

Captains who assumed the mantle of a commodore to command navy yards or squadrons at sea were expected to deal with difficulties and make decisions under duress. But what commodore other than the hapless James Armstrong had in living memory been confronted by a civil insurrection of the magnitude of the American Civil War? Not only did the machinations of secessionists outside Gosport's gates pose a threat, but Commodore McCauley faced an even more insidious foe—that of his

Southern officers "going South," first gradually by their tainted advice then *en masse*, leaving him metaphorically at the helm of an abandoned ship. Bennett offered a glimpse into McCauley's nightmare:

> All these contradictions and perplexities were too much for the commodore to unravel ... it is no wonder therefore, that the poor old man was unable to rise to the occasion. To his mind, long before narrowed to follow a straight line of naval customs and precedent, the situation was irregular and wholly inexplicable. ... Never before had he heard people talk of taking possession of a navy yard ... Never before had the majesty of a navy yard been outraged by officers walking out of the gates without leave, and without written orders endorsed by the commandant as required by regulation. And then, as if to prove that all signs had failed, the infallible regulations themselves contained not a word of instruction as to what to do in case of insurrection and the threatened seizure of the yard.[20]

In the end, Commodore McCauley was overwhelmed not so much by the external menace to his navy yard, but by the secessionists' psychological pressure and a wave of officer resignations that led to his loss of command cohesion within. He was not a perfect commandant and he had his faults. The real untold story lies not in his failures, but in how he managed to maintain control for as long as he did.

* * *

Commander Alden later quipped that he was often asked "Why didn't the commodore let you have that ship?" His answer pointed a finger at the "masked traitors" who persuaded McCauley at every turn to not move *Merrimack* for fear it would "bring on a collision with the people outside."[21] But Alden, an eyewitness to history, told only part of the truth. There were compelling and substantial reasons for Commodore McCauley's fateful decision.

First, as it has already been demonstrated, *Merrimack* had no guns on board to protect her, she had insufficient officers and crew to man, move or fight her and she hadn't yet been winded, an exacting maneuver for a big ship in confined waters. In short, she was not prepared to leave Gosport Navy Yard.

Second, obstructions were known to block the channel near Craney Island and no pilot was available to con the ship to the safety of Fortress Monroe. When later queried by the Senate committee about why he refused Alden's pleas to let *Merrimack* go, McCauley replied:

> Yes, sir; he did say that she could be taken down; but I thought differently. It is certain that she could not have got over the obstructions in the river, for there was only sixteen feet of water, and without her armament she drew twenty or twenty-one feet of water.[22]

Not reported to the committee or mentioned elsewhere at that time, was a documented episode that bore directly on the question put to McCauley. At the conclusion of her maiden voyage, the new steam frigate *Merrimack* ran aground on the morning of 9 March 1856 when steaming up the Elizabeth River and making an approach to Norfolk's naval anchorage. The weather and visibility were good, the ship was guided by two local pilots, and she was manned by her full complement of over 550 officers and crew, and yet she ran aground in broad daylight. It took twelve hours of strenuous effort to refloat her, including shifting her spar deck guns, running out an anchor astern, backing her topsails and hoisting out all her boats.[23] If the same thing had happened to the unarmed, undermanned and unrigged *Merrimack* while attempting to escape under Alden's command, the helpless frigate would have soon been captured by Norfolk's secessionist militia—if Alden didn't burn her first. Would Commodore McCauley have known about her previous grounding? Yes, if he had had any conversations with the officers around him, because *Merrimack*'s captain on that day was Garrett Pendergrast, now commodore of the Home Squadron, and one of her lieutenants, Edward Donaldson, was the receiving ship *Pennsylvania*'s lieutenant. Naval officers were acutely aware of the perils of navigating a confined waterway like the Elizabeth River.

Third, and more of a psychological barrier than an actual obstacle, was Commodore McCauley's memory of his first combat experience as a young midshipman at the Battle of Craney Island. There he learned, in the same waters *Merrimack* would have needed to traverse, that a small well-led force of determined men could seize control of Norfolk's ship channel. If its entrance could be blocked, it could also be closed to exiting vessels. The old commodore was the only officer at the Gosport Navy Yard to have had first-hand experience of the action at Craney Island, one overlooked benefit of his often-belittled abundance of years in the naval service.

Finally, the decision to move *Merrimack* was entirely Commodore McCauley's to make. Secretary Welles' last communication of 16 April plainly said "it may not be necessary … that she should leave … unless there is imminent danger pending." On the afternoon of Thursday 18 April, as McCauley saw it, the danger threatening Gosport Navy Yard was not enough to overcome the very real risks of letting *Merrimack* go.

* * *

With their mission at an end and their orders freshly endorsed by the commodore, all that remained for Commander Alden and Engineer-in-Chief Isherwood was to return to Washington and report the details of their endeavors to Secretary Welles. Two encounters they experienced on their way home illustrate how pervasive the secessionist vigilantism had become in the Norfolk area, adding further credibility to McCauley's decision to hold back *Merrimack*.

The first was an attempt to kidnap Benjamin Isherwood and hold him as a prisoner of war. So resented was the engineer-in-chief's role in restoring *Merrimack*'s machinery that a band of angry citizens planned to grab him as he left the Atlantic Hotel to catch the Baltimore boat. Only the intervention of a "warm personal friend" foiled their plot. First his trunk then Isherwood was smuggled on board and locked inside a cabin on the Bay Line steamer, and there the engineer-in-chief remained out of sight until the boat was well into the Chesapeake Bay.[24] Alden seems to have escaped notice, perhaps because his failure was too recent to draw the plotters' notice, a circumstance aided by him staying at quarters within the yard rather than at the hotel.

The second encounter was a tale related to their travel companion on the steamer to Baltimore, Gosport Navy Yard's Paymaster Edward Doran. Earlier that day, Doran realized he wouldn't be able to negotiate a New York bank draft on the usual depository in Norfolk in order to obtain the hard currency necessary to pay the yard's workmen. While Doran was in his office at the navy yard, a friendly employee tipped him off that the secessionists planned to waylay him if he attempted to carry any money from Norfolk back to the yard. He then sought out Commodore McCauley to discuss the matter. They agreed that Doran should return to Washington and request further instructions from Secretary Welles, which explains why he was traveling in the company of Alden and Isherwood.

Doran's description of his conversation with McCauley, which was held only a few minutes before Isherwood's final confrontation with the commodore, was absent of any mention of him being "stupefied, bewildered and unable to act." Rather the opposite, it portrayed the commodore as an engaged manager constantly asked to make decisions. And he was far from being "absolutely alone" that afternoon, but that would come soon enough.[25]

* * *

CHAPTER 14

Alden's Shocking Report

The three naval officers, Commander Alden, Paymaster Doran and Engineer-in-Chief Isherwood settled into their cabins for a well-deserved night's sleep. Before putting his head down, the conscientious engineer-in-chief drafted his report to Secretary Welles, relating how he succeeded in restoring *Merrimack*'s machinery but was unable to persuade Commodore McCauley to let her sail. Dated 18 April 1861, his report must have been written while he was on board the Bay Line steamer.[1] Commander Alden, who arguably had less to crow about, wrote no report that night or the next day. It wasn't until seven months later, after being chided by Secretary Welles for not submitting a report, that Alden first composed an account of what happened at Gosport.[2] This circumstance of Isherwood's written report being dated the day *before* he physically reported back to the secretary together with Alden's lapse has caused a confusion of dates in subsequent narratives—both contemporary and into the present time—of just what happened on Friday, 19 April, the day both Alden and Isherwood returned to Washington and told their stories to Secretary Welles.[3]

As the Bay Line steamer sped north up the Chesapeake Bay, the Virginia Volunteer militia under the command of General Harper invested the heights surrounding the federal arsenal at Harpers Ferry. It was only because of quick action by the army lieutenant commanding the arsenal's meager guard that Henry Wise's scheme of seizing a vast trove of arms for Virginia was frustrated. The retreating Union troops set fire to the arsenal building and stores, burning over 20,000 rifles and pistols. It was not the desired outcome, but the shops were saved, and the gun and pistol barrels and locks were salvaged, later to be worked into serviceable firearms. Better is the enemy of good enough.[4]

* * *

Commander Alden and Engineer-in-Chief Isherwood arrived in the nation's capital late on the morning of Friday 19 April. Welles later recalled the event in his diary. "My impressions are that Commander Alden called and made [sic] report on the same day with Mr. Isherwood, but he states it was on the 19th … ."[5] The secretary then lapsed into the fog previously alluded to by confusing the date of Isherwood's report with the date of his and Alden's return to Washington. This mix-up of dates pervades both the *Diary of Gideon Welles* and his serial accounts later published in *The Galaxy* under the heading "Mr. Welles in Answer to Mr. Weed."[6] The best explanation for Welles' errors is that it was not until 10 August 1862—sixteen months after the crisis that enveloped the Gosport Navy Yard—that he began making entries in his diary on a day-to-day basis.[7] In writing his accounts of the first year of the Civil War, Welles relied on his memory, notes and written reports. And he consistently mistook the date that Isherwood physically reported to him in Washington, Friday 19 April, for the date of his written report, the 18th of April. For the fast-paced events surrounding these dates, one day makes a big difference. From here on this discrepancy has been filtered out in the interests of the story line's flow.

Continuing with Welles' recollection of Alden's return:

> Commander Alden, who … had been sent on special duty to Norfolk on the 11th of April, returned on the morning of the 19th … The Cabinet was in session when he arrived. The loss of Harpers Ferry the preceding night, and the movements at Norfolk, with the threatened attack upon the navy-yard and upon Fortress Monroe, were among the matters under consideration. … Commander Alden … went to the Navy Department, and finding me absent, followed to the Executive Mansion, and, calling me from the council, related the strange condition of things at Norfolk, and the bewildered and incapacitated state of mind of Commodore McCauley.[8]

> Commander Alden … informed me, with emotion which he could not entirely suppress, that Commodore McCauley had refused to let him have the *Merrimac*, that after the fires had been kindled they had been drawn by the Commodore's command, that the vessel was at the wharf, and that the deportment and remarks of some of the younger officers left no doubt in his mind that they had control of the Commodore and the yard. … Instead of inspiring the well-intentioned but infirm old man, Alden had struck away from the yard and had immediately returned to report to the Department. I took him forthwith to the President, and the Cabinet, which was then in session, when he related what had occurred.[9]

Alden's report must have been a shock. Coming less than twenty-four hours after Commodore Paulding's unexpected return to Washington with his optimistic account of Gosport's situation and McCauley's competence, and right on the heels of the seizure of Harpers Ferry, the news could hardly have been worse. But it spiraled even further downhill as Lincoln and his cabinet came to grips with the full scope of Alden's report. Not only had McCauley declined to release the steam frigate *Merrimack* but, assuming they had understood Alden correctly, the Southern officers of the yard had staged a coup of sorts. If true, it portended the culmination of their greatest fears.

After delivering his distressing account, Alden was excused and Lincoln's cabinet meeting resumed with a renewed intensity. Before moving on to Secretary Welles' positions and the cabinet's deliberations, it is worth stepping aside to observe the protocol of what had just happened. Of the two officers who returned together from the Gosport Navy Yard, both having been dispatched for special duty by the secretary, only the line officer, Commander Alden, was introduced to the president and ushered into the cabinet meeting. Welles' recollections confirm that at some point during the day, he met with Isherwood and learned the details of his role in what happened at Norfolk.[10] And recall that Isherwood was the only one to submit a report. A reliable source, writing some years after the event, revealed how Isherwood's diligent efforts came to an unexpected conclusion:

> Mr. Isherwood made a short written report of his connection with the *Merrimac*, and the Secretary and himself never exchanged a word about it. It was tacitly understood that the subject was to be ignored, as one not politic for the public to know in the existing state of high feeling and excitement, and it was ignored.[11]

Commander Alden continued to play an active role in the Gosport Navy Yard's story; Engineer-in-Chief Isherwood's involvement, at least as far as its public view, was finished.

* * *

Now having the cabinet's full attention, Secretary Welles launched into a summary of what was at stake at Norfolk. First, he recommended that immediate steps be taken to protect the considerable amount of public property there—ships, ordnance, stores and tools—all of which were now in jeopardy. Second, from both a naval and a military perspective, it was a necessity to retain control of the Gosport Navy Yard. With its graving dock, extensive workshops and now being the southern-most of the navy yards, it was irreplaceable. And third, the consequences of the yard's being abandoned would be disastrous—just think of what mischief the secessionists could get up to using its facilities. However, to protect the public property and prevent a foreseeable calamity, a military force needed to be rapidly dispatched to protect the navy yard. And for that, the cooperation of the Army's Lieutenant General Winfield Scott was vital. But from his previous conversations with the old general, Welles knew Scott was firmly opposed to sending any soldiers.[12]

After hearing out Secretary Welles, Lincoln and the cabinet concurred with his conclusions and urged that he and the president immediately go and visit General Scott to see what could be done. This same day the Lincoln administration proclaimed a blockade of the coast of the Confederate States, ranging from South Carolina to the Rio Grande, a wildly ambitious undertaking considering the overwhelmed state of the U.S. Navy. And later that afternoon, the telegraph would begin to buzz with the startling news of riots and bloodshed in Baltimore as the first regiments of newly called up Pennsylvania and Massachusetts volunteers marched through that city on their way to Washington.[13] It was shaping up to be a turbulent and fateful Friday, unlike any other experienced by Lincoln and his government.

The president and his navy secretary then walked from the White House across to General Scott's military headquarters in the nearby War Department building. There they met with "Old Fuss and Feathers" and Welles quickly recapitulated his views on the fast-changing situation in Norfolk and the insurgent threat to a vulnerable navy yard that was wholly without military support. Secretary Welles reminded the general that the Gosport Navy Yard was unfortified and but weakly defended. The small force of sailors and marines could not be expected to hold out for long. Further, the Department had just received the unsettling news that vessels were being sunk to further block the channel.[14]

After hearing him out General Scott reiterated his former position. He was adamantly against sending any of his precious regulars away from Washington to guard the navy yard or for any other purpose. Scott pointed out that Washington, a Southern city, was entirely surrounded by slave-holding states. Waving his hand at a map, he outlined how exposed the nation's capital was. As the general put it, they "were asking for an impossibility." If he didn't have the soldiers to save the federal arsenal at Harpers Ferry, just up the Potomac, where was he to find the men to send deep into Virginia's "enemy's territory"?[15]

Then the general stabbed his finger at Norfolk and Tidewater Virginia. Even if he had the men to send to the Gosport Navy Yard, the place was indefensible. Any force sent there was sure to be captured. And the government could ill-afford that. Maybe more naval personnel and a detachment of marines could be sent; if the worst happened the Navy might succeed in evacuating them with its ships—that is if the channel wasn't fully blocked. But there would be no escape for his soldiers. Here the general paused to remind Lincoln that just a few days before, he had urged that spare marines from the Philadelphia and Boston Navy Yards be dispatched to Gosport. The arrival of such a relief force, he suggested, would cause the secessionists to enlarge their preparations to threaten Norfolk, thereby buying time for the North to send volunteer regiments to defend it.[16]

Shifting his focus across the waters of Hampton Roads, Scott told Lincoln and Welles that even the mighty bastion of Fortress Monroe wasn't sufficiently garrisoned to repel the rebel forces that were, according to his best information, preparing to attack it. And then, returning to Washington, the general reminded them again that he did not even have enough men to defend the nation's capital should a "formidable demonstration" be made against it. General Scott summed up his difficult position:

> For the present, his first great duty, with his feeble force, was to defend Washington, and next to Washington, Fortress Monroe, which was the key to Washington, Norfolk, Baltimore, Chesapeake Bay, and the rivers which entered it. He therefore could not, and would not, consent to part with a single regular for either Harpers Ferry or the Norfolk Navy Yard; and ... that public property in each of those places must, in the case of an attack, be sacrificed.[17]

While Scott's view of the situation must have been uncomfortable for the president to hear, one cannot help but admire General Scott for delivering such a candid summary to his commander-in-chief. But evidently Lincoln and Welles must have made some impression on the old general for he finally relented and agreed to allow Colonel Delafield from the engineer corps to assist with strengthening the navy yard's defenses. By this moment in time, Welles had already set his mind on dispatching a substantial expedition to confront the secessionist threat to Gosport and retrieve as many of the Navy's vessels as possible. But such an expedition needed more than just sailors, marines and an army engineer, it desperately needed the teeth that only a military force could provide. And so, after more pressure from Lincoln and Welles, General Scott, at last, consented to allow a battalion of Massachusetts volunteers, assuming they reached Fortress Monroe in time, to accompany Welles' naval expedition. But, the old general sternly warned, "they were undisciplined and would be good for nothing ... but ... in throwing up batteries under the direction of an engineer."[18]

* * *

Welles now had a commitment for a military force—a mob of untrained militia. But it was all that could be spared and it would have to do. Now to finalize the myriad moving parts of his expedition. As he later wrote "I lost not a moment, after parting from the President and General Scott in giving the following order to Commodore Paulding:"

> You are directed to proceed forthwith to Norfolk and take command of all the naval forces there afloat.
>
> With the means placed at your command, you will do all in your power to protect and place beyond danger the vessels and property belonging to the United States. On no account should the arms and munitions be permitted to fall into the hands of insurrectionists, or those who would wrest them from the custody of the Government; and should it finally become necessary, you will, in order to prevent that result, destroy the property.
>
> In carrying into effect these orders, you are invested with full powers to command the services of the entire naval force in carrying out these instructions. It is understood that the War Department will detail Colonel Richard Delafield, or some other competent officer with a command, to aid and assist in protecting and guarding the yard and property at Gosport and vicinity, and you will cooperate with that officer in this object.[19]

For very understandable reasons, Secretary Welles chose Commodore Hiram Paulding to lead his new expedition, a bold undertaking of near existential importance to the federal government. Paulding was available and familiar with the situation in Gosport, having returned only the day before. But in leaving Norfolk so soon, without having grasped the nature of the imminent threats to Gosport, Paulding, a man as old and careworn as Commodore McCauley, demonstrated a negligent streak that should have given Secretary Welles pause and motivated him to select someone better able to perform such a vital service. History hinges on such decisions. Had Welles chosen a more competent officer, Gosport Navy Yard may well have experienced a different fate.[20]

A curious aspect of Paulding's dispatch is the date of Welles' orders, Thursday 18 April 1861. This was the day that Paulding returned to

Washington and when Welles must have realized he needed to send the commodore back to Gosport, thus initiating preparations for an expedition and giving the nod to Paulding that he should be prepared to go. And Welles did issue a flurry of other orders to reinforce Norfolk on that same day. But Welles' *written* orders to Paulding clearly referenced the inclusion in the expedition of "Colonel Richard Delafield …with a command," a concession apparently granted by Scott following the receipt of Alden's shocking report in Washington, the involvement of the president and his cabinet and the subsequent meeting of Lincoln and Welles with the general, all of which happened on 19 April. In two separate reports by Paulding, the later written no more than two weeks after Welles issued his order, Paulding referred to "your order of the 19th instant"[21] and wrote that he "was ordered on Friday the 19th ultimo to take the war steamer *Pawnee* …"[22] And finally, Paulding tells in his biography "I … left for Norfolk on the evening of the day on which my orders were given,"[23]

Putting aside any questions about their date, Paulding's orders were but one of a steady stream of instructions that flew from Welles' desk. First and foremost, it was necessary to get Paulding and his command to Norfolk and at that moment, there was only one naval steamer available at the Washington Navy Yard, the tender *Anacostia*. This humble former merchant tug mounted but one gun was 129 ft long and could at best make seven knots.[24] She was assigned to Commodore Paulding.

Next came orders to officers at the Washington Navy Yard for placing gun powder and other combustible material on board *Anacostia* to give Paulding the option to destroy the navy yard and any ships or matériel that could not be removed. The necessary instructions were sent to the yard's commandant, Commodore Franklin Buchanan and its ordnance officer, Commander John Dahlgren.[25] The emphasis on combustibles at the expedition's planning stage yields unmistakable evidence of the Navy's intent to destroy assets it could not recover. Dahlgren later tallied the combustibles he loaded out as forty barrels of gunpowder, eleven tanks of turpentine, twelve barrels of cotton waste and 181 portfires (slow-burning fuses).[26]

At this point, Commodore Paulding unexpectedly put his foot down. He informed Secretary Welles that he "declined" to sail in the vessel assigned to him, *Anacostia*, "a miserable tug" that was "without capacity or any quality … necessary for the service I was to perform."[27] The commodore's belligerence was empowered by the discovery that another steamer, this one "efficient and ably commanded," had just hove into sight on the Potomac. At 8:45 that morning the steam sloop *Pawnee*, returning from her unsuccessful mission in support of Fox's Fort Sumter relief expedition, cast her anchor off the entrance to the Washington Navy Yard.[28] *Pawnee* was everything *Anacostia* was not: she was large, roomy, fast and well-armed. Further, she was the U.S. Navy's only war vessel to carry her 1,500-ton displacement on the light draft of 10 ft and, even better, she was propelled by twin screws and capable of steaming at ten knots.[29] Of all the Navy's vessels at Secretary Welles's disposal, none was better suited than *Pawnee* for a rescue mission demanding a vessel that could accommodate extra officers and men, deftly skirt obstructions in a river channel and tow a warship to safety. But first Paulding had to bend Secretary Welles to his will. *Pawnee*, the only effective warship on the Potomac, was dearly coveted for the defense of the capital. After "earnest solicitation and the promise to return with her ... at the earliest practical moment" the commodore finally won over Welles.[30] Paulding got his efficient steamer.

* * *

Now followed even more expedition force orders. The first was to Colonel Harris of the U.S. Marines, and it instructed him to detail one hundred marines, together with suitable officers, for immediate service on *Pawnee*.[31] Welles' decision, once *Pawnee* sailed, would leave the Washington Navy Yard defenseless at just the moment when most of its senior officers chose to resign and "go South." His next two orders were directed to the captain of the newly reassigned steamer *Pawnee*, Commander Stephen Rowan, the first notifying him his ship must be prepared to sail without delay and that Commodore Paulding would soon join him with full instructions. The second order informed Rowan that he could expect Army Captain Horatio Wright—Colonel Delafield's substitute—and that he should be comfortably accommodated.[32] The frenzied state on board *Pawnee*, with her normal ship's service needs for coal, water, stores and sundry items, all to be accomplished while preparing for immediate departure, can only be imagined.

Welles and Paulding had not neglected the vital necessity of organizing a group of naval officers to command any ships at Gosport that could be saved. In this work they were greatly assisted by Paulding's position as head of the Office of Detail—he had at his fingertips the status of every active U.S. Navy officer.[33] That they were able to select, order for duty and assemble the thirteen officers that joined *Pawnee* before she sailed Friday evening is little short of astonishing.[34] Perhaps if such energy and diligence had been focused a week before on Gosport Navy Yard's needs, Paulding's expedition wouldn't have been necessary. But Secretary Welles had only been at his job for forty-five days, so a charitable construction is that he was a fast learner.

Familiar names populate the tally of one captain, four commanders, seven lieutenants and a chief engineer who accompanied Commodore Paulding. Topping the list was Captain Charles Wilkes assigned to command *Merrimack*, the "stormy petrel" and leader twenty years before of the U.S. Exploring Expedition; Commander Rodgers detailed to *Plymouth*, one of the future Admiral Du Pont's "clique" during his ironclad assaults off Charleston; Lieutenant Henry A. Wise—a New Yorker not Virginia's ex-governor—an ordnance officer who would later become one of Welles' bureau chiefs; and finally a familiar face, Commander James Alden who was assigned to take charge of the brig *Dolphin*, something of a demotion from his last excursion to Gosport.[35]

As for the crews to be apportioned under these thirteen officers, Captain Wilkes' mission report outlined the expedition's numbers: *Merrimack*, 50; *Germantown*, 40; *Plymouth*, 30; *Dolphin*; 20; "besides the marines to be detailed." [36] The expedition's assignments of officers and men provide a ready measure of just how short Commodore McCauley had been of resources. But as for mustering the crews, they were simply not to be had in Washington. Instead, Paulding says that he "sent to New York for the dispatch of seamen."[37] What response did he receive after weeks of false hopes? That same day Commodore Breese advised the secretary "the draft ordered to Norfolk will leave tomorrow, probably on the regular Norfolk steamer at 3 p.m.," and added that some 173 recruits were currently available and more should follow. That meant, if all things went well, the crews would arrive in Norfolk on Sunday 21 April.[38] But the news from Baltimore, where rioting had just erupted over militia transits, was not good. Commander Hunter was obliged to report that Baltimore's hometown steamship company, the Bay Line, declined to take naval recruits for passage to Norfolk.[39]

By Friday evening, Secretary Welles and Commodore Paulding had performed a miracle of improvisation. They lashed together out of thin air a significant fighting force of 100 marines, fourteen experienced naval officers including Paulding, which together with *Pawnee*'s compliment of some 120 officers and men and the nearly 200 seamen recruits to be dispatched from New York, would give the Navy the muscle to reassert its authority at Gosport and permit the orderly retrieval of its valuable assets. But the one grievous omission of Secretary Welles—"an unforgivable error" according to one historian—was to *not* send word to Commodore McCauley that help was on the way.[40] He had the means to do so, either from Fortress Monroe which was still in easy communication with Washington, or more directly, by the dispatch of an officer on a Bay Line steamer. Despite Lincoln's declaration of a blockade of the Confederacy,

Bay Line boats maintained their normal scheduled service to Norfolk until the end of April.[41] Perhaps Welles was afraid that if the word got out about Paulding's expedition, the secessionists would move in force on the navy yard, but as will be shown, if McCauley had known that relief was soon at hand the crisis that enveloped the Gosport Navy Yard would have had a far different ending.

That Friday evening *Pawnee* slipped her moorings off the arsenal wharf at Greenleaf Point and headed down the Potomac River.[42] At her best speed, she would arrive at Fortress Monroe, 132 miles distant, on mid-afternoon of the next day. The twelve-knot Bay Line steamer that left Baltimore that same evening was scheduled to arrive at just past 6 a.m. the next morning. For a navy yard under siege, eight hours is a very long time.

* * *

CHAPTER 15

Norfolk Showdown

The newly independent Commonwealth of Virginia's first acts of aggression toward the federal government came soon after the passage of its ordinance of secession. On Thursday, 18 April 1861, Governor John Letcher telegraphed orders to General Harper at Staunton, instructing him to march his volunteer companies to Harpers Ferry and seize the arsenal, and he issued orders to Major General William Taliaferro, sending him to Norfolk to assume command of troops ordered into service from that district. Both actions, although officially sanctioned by Virginia, were the brainchild of its former governor Henry Wise.[1]

General Taliaferro arrived in Norfolk early the following day, the same morning that Commander Alden and Engineer-in-Chief Isherwood unexpectedly reappeared in Washington seeking Secretary Welles. Taliaferro was accompanied by staff officers, Colonel Henry Heth, aide-de-camp, and Major Nathaniel Tyler, and also by Captain Robert Pegram representing Virginia's provisional navy. Pegram and a fellow officer, Captain Catesby ap R. Jones, had been authorized to take command of Norfolk's naval station and organize its naval defenses.[2]

William Booth Taliaferro was no stranger to Virginia's militia, its politics, or to the U.S. Army. Scion of an old family with roots in Gloucester County stretching back to 1655—they pronounced their surname as "TOL-i-fer"—the thirty-eight-year-old general had served in Virginia's House of Delegates, practiced law and dutifully managed his plantations. He rose to the rank of major in the U.S. Army during the Mexican War and, serving as a major general in Virginia's militia, commanded the volunteer companies mustered for guard duty during John Brown's trial and execution. That experience welded his friendship with Governor Henry Wise. Taliaferro was also the nephew of James Seddon, who would become Jefferson Davis' secretary of war.[3]

General Taliaferro found Norfolk in a "state of great excitement" and such was the extent of unrest that the governance of the city's civil affairs had been "committed to the management of a committee of safety, composed of the most respectable and intelligent citizens." The general quickly discovered the cause of Norfolk's excitement: hostile actions recently taken to obstruct "the channel of the river some 9 miles below the city, thus attempting to prevent of egress of the vessels of war from the harbor." Taliaferro's point of blockage maps out to just opposite Sewell's Point. He was a witness to Virginia's largest seaport holding its breath in anticipation of United States forces reacting to its intelligent citizens' scheme to paralyze the Gosport Navy Yard.[4]

The general and his staff took stock of the local military resources at Virginia's disposal. Taliaferro's count conforms with the known number of militia companies serving in the Tidewater area, which he described as:

> a battalion of volunteers and a few detached companies, in all numbering not a great deal over 300 men, and the force of the town of Portsmouth to consist of a regiment of volunteers of about the same strength. The militia of the two towns was without arms and there was no naval force in the State service to cooperate with me. The only artillery consisted of a few 6-pounder field pieces.[5]

Perhaps even more worrisome was Norfolk's state of fortifications and defenses—the harbor and its approaches were entirely unprotected. The old works at Craney Island and Fort Norfolk were in disrepair. Even if fortifications had been erected, no heavy guns were available to mount behind their earthworks. Tidewater Virginia was awakening from the long slumber of peace and, now on the edge of civil war, found itself wholly unprepared. On the other hand, Taliaferro was impressed with the forces that guarded the one remaining node of federal authority, the Gosport Navy Yard. He recognized, like many others had before, that the yard's perimeter defenses were insignificant: "only a small force of marines … protected by a trifling brick wall." But he astutely identified the Navy's "ships-of-war stretched out along the river, their heavy batteries bearing upon all parts of it," as the core of the yard's defensive power. Taliaferro's principal concern was the sloop *Cumberland* but also included the brig *Dolphin* and the receiving ship *Pennsylvania*. An experienced military man, the general understood that because these floating batteries "had springs on their cables with which to maneuver, had netting with which to resist boarding, and were using every precaution of guard boats, sentries, etc. to prevent surprise," they were virtually impregnable to waterborne assault. Estimates submitted to Taliaferro by Captain Pegram, late of the

U.S. Navy and presumably familiar with Gosport, projected that a minimum of 5,000 men would be required to seize the yard in this way. As he reasoned in his report to the governor, the option of taking the navy yard's guard ships by boarding was untenable. The only other realistic alternative was to assemble heavy ordinance behind newly erected batteries, which would then enable him to enforce demands for the yard's surrender. [6] But General Taliaferro's favored alternative required patience and more resources, so he immediately began issuing orders and urging actions aimed at swiftly nullifying the threat of McCauley's great guns.

* * *

Meanwhile, inside the Gosport Navy Yard, Commodore McCauley took little comfort in the realization by others that his navy yard was strongly protected. Rather the opposite. The cumulative effects of his increased isolation, the innumerable acts of local hostility and a gnawing fear of the unknown worked to cloud his judgment. On the morning of Friday, 19 April, the yard's civilian employees absented themselves from their daily muster. The eerie silence outside McCauley's commandant's office spoke to the derangement of affairs at the navy yard. It had also been decided that day, because of the growing danger of violence, that naval officers would not be allowed to make their usual visits to nearby Norfolk. The cutting off of communications by mail, telegraph and the flow of daily gossip fueled the outlook that they were a military force that had been surrounded—a siege mentality.

As Commodore McCauley glumly reflected on his fast-changing situation, he must have wondered how much longer he could hold out. Since Virginia passed its ordinance of secession, the trickle of resignations by his Southern officers had morphed into a virtual flight from duty, astonishing the grey-haired commodore with both its speed and lack of decorum. One day they were attending to their duties and the next they were gone, all without his permission and certainly not in accordance with naval regulations. Commodore McCauley must have felt in varying degrees anger, chagrin and mortification.

It is difficult to know at this remove the precise day-to-day, hour-by-hour progression of Gosport's Southern officers abandoning their posts, but a reasonably accurate picture can be sketched out for the afternoon of Friday 19 April. Of McCauley's eleven officers managing direct navy yard functions, only three remained on duty, Lieutenant Alexander Semmes, the yard's sailmaker and Paymaster Doran, who was away in Washington. The rest, eight in all, had vanished. The loyal Lieutenant Semmes, a relative of a man who would become infamous, Raphael Semmes of the Confederate cruiser *Alabama*, had been assigned to the brig *Dolphin* as her acting commander. Although out of commission, she was armed, lightly manned and positioned to guard the yard's upper reaches. The near decapitation of Gosport's executive staff was offset that day by the arrival of Commander Robb's replacement, Commander John Livingston. While very welcome, Livingston's presence did little in the short term to offset the loss of McCauley's experienced staff.

Officers in charge of Gosport's ordnance, ordinary and rendezvous functions had to a man disappeared, six in all including the personable Commander Richard Page who had so forcefully pleaded with Commodore Paulding for the yard's Southern officers' relief. Perhaps most crippling and, as will soon be seen, lethal to the safety of the navy yard was the disappearance of all of its ordnance officers: Commander Tucker, Lieutenant Sinclair and Gunner Oliver. Thus, the yard's most valuable assets next to its warships—ordnance, powder, shells and their irreplaceable accessories—were all left unsecured.

The receiving ship *Pennsylvania*'s 120-gun majesty was severely impaired by the resignation of her captain, Commander Sinclair, executive officer, Lieutenant Poindexter and marine officer, Captain Rich. Of the three loyal officers left on board, only Lieutenant Donaldson was a line officer, the others being her surgeon and sailmaker.

The yard's marine contingent had not escaped the carnage. First Lieutenant Baker resigned, leaving Lieutenant Colonel Edelin the sole marine officer at the navy yard to oversee its eighty-three marines. Looking at the total picture, of the twenty-nine officers who had been on duty at the yard when Virginia announced its fateful decision, two days later only eight remained at their posts.[7] A pair of replacements recently detailed to Gosport by Secretary Welles would report for duty over the next days, but being new and unfamiliar with the navy yard, their appearance gave little more than a psychological boost. But in Gosport's shattered, disorganized state, any help was dearly welcomed.

* * *

Although his staff was severely diminished, Commodore McCauley somehow managed to maintain a modicum of order over Gosport Navy Yard's 108-acre complex of buildings, workshops and wharves. His first line of defense, the yard's marine detachment, was under arms on guard duties, manning the yard's gates and on patrol throughout the night. Further, Lieutenant Colonel Edelin had been ordered to drill his men with their 6-pounder howitzers, which were kept ready for use at all times. Although the navy yard was enclosed by a substantial brick wall, ten feet high and eighteen inches thick, it offered but scant protection. Both Edelin and McCauley knew that nothing but a large force could adequately hold the yard against a determined assault by land: its grounds presented a very unfavorable field of battle, with few naturally defensible positions and many covers for hostile troops.

Beyond the yard's grounds, McCauley saw to the coordination with Commodore Pendergrast's flagship, *Cumberland*, which maintained a full watch at her guns, boarding nets stretched along her sides and a guard boat circling the anchorage. At that time the Gosport Navy Yard had 179 seamen and landsmen enrolled for service. Most were on board the receiving ship *Pennsylvania*, but some had been detailed to support Lieutenant Semmes on *Dolphin*, while others kept watch on the silent hulks in ordinary or helped secure the yard. Despite McCauley's loss of so many officers, his enemies were presented with the façade of a commandant who had taken every possible precaution to defend his navy yard.[8]

While the commodore struggled to preserve Gosport Navy Yard's fragile dominion, his enemies feverishly schemed to spread every rumor, create deceptions and encourage outright falsehoods, all to convince him that he was trapped and his best recourse lay in an honorable surrender. The near daily routine of towing an old schooner down the Elizabeth River to be sunk off Sewell's Point was joined by another scare tactic, this one implemented by a bright and loyal Virginian serving as president of the Norfolk & Petersburg Railroad, William Mahone. A thirty-five-year-old graduate of Virginia Military Institute and later known by his military title of general, Mahone devised a very effective stratagem to create the illusion of the arrival at Norfolk of endless numbers of trains full of troops.[9]

To fully appreciate the deceit behind Mahone's plan, an understanding of the geography of Gosport Navy Yard's eastern boundary is helpful (see Map No. 9). Because of the branching of the Elizabeth River and the water barriers it created, the Norfolk & Petersburg Railroad approached Norfolk from well south of the Gosport Navy Yard, where it spanned the river's southern branch at a relatively narrow crossing point. From there, the railroad turned north and ran parallel to the river about three-quarters of a mile east of the water's edge. In doing so it passed close by the navy yard's forty-acre St. Helena annex used for the storage and repair of ordnance, also on the east bank of the Elizabeth. The railroad terminated in Norfolk at a station on the bank of the Elizabeth's eastern branch. This waterside station was in a line-of-sight to vessels lying at the entrance of the Gosport Navy Yard—it could be seen from *Cumberland*. With the map tour in place, Mahone's deception is readily grasped:

> Trains of cars were sent out empty from the station at the waterside, and at a short distance from the city were filled with troops which had gone out for the purpose, and were then brought back, landing the men in plain sight of the ships, and this was repeated many times a day, of course conveying the impression that fresh troops were constantly arriving.[10]

Other accounts tell that the incoming troops were encouraged to yell, hoot and shout, which could be heard by observers at the navy yard. This *ruse de guerre* together with the stoking of endless rumors of the imminent arrival of Southern troops steadily worked on the minds of McCauley and his command, inducing the fear that the secessionist menace was growing.

Another factor amplifying the effects of Mahone's deception was that it drew attention to Gosport Navy Yard's Achilles heel, its exposed St. Helena annex. The Senate committee hearings provided a thumbnail sketch of the surrounding terrain: "There are no hills in or about Norfolk commanding the yard, but the country in the vicinity presents a flat, champaign surface."[11] Thus, a surrounding enemy force who wished to seize the navy yard intact and who also understood that the key to its defenses was the men-of-war guarding it, would soon conclude that the optimal location for a battery to command and potentially destroy the guard ships was the unprotected St. Helena annex. Lying across the Elizabeth River from the main yard, shrouded by thick stands of pine trees and accessible from the Norfolk & Petersburg Railroad, it was the perfect place for the enemy to stealthily site his guns. And that is just what General Taliaferro concluded: he had already telegraphed a request for heavy guns to be sent to "arm a field work I had projected at a point called St. Helena, commanding the anchorage and navy yard."[12]

What Taliaferro could see, so could Commodore McCauley. Earthworks appearing at St. Helena were his greatest nightmare; with his scarce resources he could threaten them but he couldn't stop their emplacement. Hearing Mahone's trains rumble by all day long had begun to play on his mind. McCauley later testified:

> I heard that two thousand men came down Friday, the 19th of April, from Richmond, Petersburg and other places in the vicinity of Norfolk. I was then satisfied that I had better get the ships out of their way. But it was my intention to remain and defend the yard.[13]

* * *

So far all of Commodore McCauley's actions had been mostly passive, steps taken to ensure a more secure mode of defense. But by Friday evening he decided to take things one step further and prepare for the worst. The time had come to destroy property that might be useful to the insurgents if it became necessary to abandon the yard. Accordingly, the order was given to spike all the guns lying in the yard's numerous gun parks, 1,200 in all (see Appendix 5). "Spiking" a naval gun of that era involved hammering a metal object—the spike—into the gun's vent opening, which made it impossible to fire the weapon until the obstruction could be drilled out and the vent dressed. Maiming a large caliber gun in such a way was hard work and parties were mustered under Lieutenant Semmes, joined by Marine Captain Kintzing from *Cumberland*, to begin the sad job. Later reports indicated that the majority of the guns had been spiked with common wrought iron nails which proved largely ineffective; the most seriously disabled weapons were those spiked with rat-tail files.[14]

Simultaneous with the mass spiking operation, large quantities of ordnance stores were crammed alongside the mass of small arms already loaded on *Cumberland*, rendering her a veritable "floating magazine," greatly increasing the danger to her officers and crew should hostile fire be encountered. What couldn't be spiked or carried off was destroyed and dumped in the river. Included in this "abominable vandalism" were hundreds of small arms and thousands of shells. If it couldn't be removed or broken, it was heaved into the water.[15]

As dusk settled over Tidewater Virginia, the secessionists began their own version of vandalizing federal property. Their target: Fort Norfolk, the Navy's arsenal holding its store of gunpower. Here again, the evil of Gosport's wave of officer resignations showed its face—the yard's former ordnance lieutenant, George "Terry" Sinclair, now a captain in Virginia's provisional navy, masterminded the arsenal's seizure. One can imagine him opening the gate with his own keys. And this was no casual raid snatching a wagon full of booty, but a sustained, well-planned effort that required the impressment of a coastal steamer, lighters and small craft; of "all the carts and horses that could be procured;" and which was protected by volunteer companies of militia and artillery, supported by two 6-pounder howitzers. After securing and surrounding Fort Norfolk, the secessionist forces removed 1,300 barrels of gunpower, loaded them on lighters and small craft and sent them on their way up the James River to Richmond. The "residue" consisting of 1,500 barrels of gunpowder was removed to "a point beyond the range of the guns from the water," or outside the reach of U.S. Navy retribution. That one night's work netted Virginia's militia forces some 125 tons of gunpower, all of it removed by untrained volunteers at considerable risk to life and limb. But as the sun rose on Saturday, 20 April 1861, their job was largely done.[16] Success had crowned their bold action.

* * *

Once word of Fort Norfolk's seizure filtered into the Gosport Navy Yard, Commodores McCauley and Pendergrast must have been appalled. Not only had the Navy's store of gunpowder been lost, but far worse, the enemy possessed it. It was only a matter of time before heavy guns appeared and now the rebels had all the powder they needed to rain deadly fire on *Cumberland* and every other ship in the stream.

All too soon their worst fears materialized. A breathless Lieutenant Selfridge urgently knocked on the commodore's door and reported that *Cumberland*'s masthead lookout had spotted earthworks being constructed behind the woods of the St. Helena annex. Selfridge's memoirs dryly record McCauley was "much perturbed at the news." Horrified and shocked might be a better description. The commodore then asked Selfridge for his opinion on what should be done. The brash young lieutenant unhesitatingly proposed he be sent to Norfolk and firmly impress on their commanding general "that if the work of mounting the batteries continued, the *Cumberland* and other vessels present would fire on the city."[17] The message McCauley finally settled on was rather less drastic: should the secessionists persist in menacing him by placing batteries opposite the yard, "I would consider it an act of war and fire upon them."[18]

Before the hour passed, Lieutenant Selfridge had been pulled across Norfolk's harbor under a flag of truce. A loud and excited crowd greeted him at the dock. He was fortunate to be spotted by a familiar face—Captain Pegram of Virginia's provisional navy—who was acting in the capacity of a provost marshal. The career trajectories of these two officers yield a fascinating glimpse into the antebellum Navy. Robert Pegram won his lieutenant's commission in 1841 and still held that rank twenty years later when he resigned to "go South," while the younger Selfridge had worn his lieutenant's star for little more than a year.[19] One wonders how the lieutenants greeted each other.

Selfridge was escorted to General Taliaferro's headquarters, where in "forceful terms" he delivered the commodore's message. The general shot back, saying that firing on an undefended city would be a "terrible thing," his words leaving doubt as to whose message Selfridge actually delivered. The lieutenant replied that "it would also be a terrible thing ... if the yard and men-of-war were destroyed." From there the interview dissolved into small talk and finally ended with the agreement that Taliaferro's aide-de-camp, Colonel Heth, would accompany Selfridge back to the navy yard and meet with Commodore McCauley to further discuss the matter.[20] The commodore later gave his recollection of the meeting's outcome:

> I sent Lieutenant Selfridge, but the town was in such a ferment that he could obtain no definite information, further than the inferences which he drew from the numbers of troops he saw passing and repassing, and the fact which fell under his notice that troops were coming in all the time in the cars.[21]

To the modern-day observer, it seems obvious that Taliaferro and his officers were playing Selfridge and McCauley for all they were worth, telling them nothing, parading their endlessly recycled troops and conniving to get Heth into the yard to gain first-hand intelligence about the situation inside Gosport. Their charade carried on well after Selfridge introduced Heth to the commodore. The colonel spent "much time in discussing a variety of matters, but refrained from bringing up the question of the earthworks." It took a private word from Selfridge to McCauley to refocus the conversation on the main issue, where once restated, Colonel Heth gave assurances that the work would cease.[22]

It was easy for the insurgents to give fulsome assurances about something that they weren't yet doing and which they now knew their federal opponents greatly feared. During those tense days in April, no earthworks were being dug by Taliaferro's men nor were any batteries being emplaced at St. Helena. The general was waiting for assurances that heavy guns were on their way to him before risking the exposure of his working parties to naval bombardment. All accounts confirmed this, especially testimony later given by navy yard mechanic, James Clements. The loyal Clements, who "went to Norfolk ... and went all about that region," chased down the earthwork rumors and made inquiries of his friends, "but could see no signs of any battery." He attempted to report his findings to McCauley, but getting to the commodore inside the hot-house atmosphere of the Gosport Navy Yard proved impossible.[23] Commodore McCauley had become swept up in the latest crisis to unfold at the navy yard.

* * *

CHAPTER 16

McCauley's Scuttling

By mid-Saturday morning the tension gripping Gosport was palpable to all within its walled grounds. A few faithful employees still labored in its workshops, but they were a mere shadow of the yard's normal bustling workforce. Marines under arms patrolled the grounds, tense and wary in their unaccustomed role, casting glances at the novel sight of their howitzers gleaming and at the ready. Beyond the yard's gates, the shady streets of Portsmouth echoed with rumors that rebel batteries were being erected across the river at St. Helena.[1] Even more sinister, word had spread that the Navy was preparing to burn the yard which if true, meant any homes outside its walls were at risk of a devastating fire storm.[2] And by now the residents of Portsmouth, Norfolk and the surrounding countryside had heard that Fort Norfolk's arsenal had been plundered. It could hardly have been kept secret. The region's whole volunteer force, nearly a thousand men, had spent all night long laboring at their rebellious deed and their stories had quickly spread.

But the hotheaded insurrectionists' work was far from done. For days they had fixed their sights on the steam tug *Yankee*—her very name made their blood boil. Lying in the shadow of the sloop-of-war *Cumberland* since her unexpected arrival in Norfolk, a refugee from Gustavus Fox's attempt to relieve Fort Sumter, the tug served as a visceral symbol of the federal government's willingness to trample Southern states' rights and promote the agenda of that villainous Black Republican, Abraham Lincoln. And steamers were desperately needed by Virginia's provisional navy. At eleven o'clock that morning they made their move. Commodore Pendergrast, whose flagship was a witness to the event, later described what happened:

> The rebels … attempted to take the government tug *Yankee* from the navy yard. The surveyor of the port of Norfolk came over for that purpose from the custom-house in another small steamer,

> with forty or fifty armed men, to take the tug, but was prevented by Commodore McCauley calling out the marines and ordering them not to give her up. … It was done under a pretense of the violation of some custom-house regulation on the part of *Yankee*.[3]

As soon as the approaching tug's intentions were unveiled, *Cumberland* immediately beat to quarters and trained her forward 10-inch pivot gun on the secessionists.[4] One of her marines, Private Daniel O'Connor, witnessed the confrontation and wrote about it in a letter to his sister:

> They tried to take the *Yankee* … but they were in dread of our long tom on the forecastle … they let her go very quick when we pointed our pivot gun at her. They spoke of massacreeing [sic] the whole of us & taking the navy yard that night & shipping but we intended to sell our lives as dear as possible.[5]

The insurgents' attempt to grab *Yankee* had been firmly deflected by McCauley rallying his marine contingent and by *Cumberland*'s decisive reaction, but as Commodore Pendergrast later testified, it "was the first act that looked to be the commencement of hostilities."[6] That Commodore McCauley also saw the secessionists' aggression as a turning point was later confirmed: in the only instance where he cross-examined a witness during the Senate committee hearings into the loss of Gosport, his opening question served to demonstrate he had acted promptly to protect *Yankee*.[7]

Yankee's attempted seizure raises a host of unanswered questions about why Commodore Pendergrast, flag officer of the home squadron, didn't employ her to command the Elizabeth River and actively bolster the navy yard's defenses. She was a good-sized coastal steamer, already on government charter and fully manned. She later actively served the Navy as an armed tug. The unrealized potential of the steam tug *Yankee* is one of the great mysteries of the Gosport Navy Yard saga. Writing in the years following the Civil War, Admiral David Porter, the apotheosis of a fighting naval officer, offered his thoughts on how a vessel like *Yankee* could have materially altered the balance during the showdown at Norfolk:

> One tug with a twenty four pound howitzer on board, properly served, would have been master of the situation, and if the Navy Department had displayed a little forethought in this emergency, the government would have been saved deep humiliation and a loss in ships, guns and stores not easy to repair.[8]

An armed *Yankee* could have dominated the Elizabeth River. And the rebels had nothing of substance to oppose her. Once the Pandora's box of her potential is opened, the possibilities bubble up. She could have stopped the sinking of light-boats in the channel off Sewell's Point. She could have disrupted the seizure of Fort Norfolk's arsenal and prevented insurgent steamers and lighters from taking its gunpowder to Richmond. She could have held hostage the troop trains crossing the Elizabeth River's Eastern branch into Norfolk's waterside station. She could have carried urgent summons for help to Fortress Monroe, less than fifteen miles distant. In the end, she more than proved her worth by towing *Cumberland* out of the burning navy yard and hauling her across the channel's obstructions.

One speculative answer to this conundrum is that perhaps Commodore Pendergrast *did* grasp the value of *Yankee*, but held her too closely to him as his command's ultimate escape capsule.[9] *Cumberland* was a sailing vessel, with all the liabilities that conferred when under way in confined waters. A powerful battery while stationary on her springs, she would have been at great risk when moving under sail in the Elizabeth's narrow reaches. If the rebels had closed in on Gosport and begun bombarding *Cumberland*, Pendergrast would have been forced to abandon the sloop to avoid her being turned into a charnel house—which is exactly what happened a year later when *Cumberland* was destroyed in Hampton Roads by the resurrected *Merrimack*. Private O'Connor revealed what *Cumberland*'s crew had in mind Saturday morning after the phantom sighting of the earthworks' construction behind St. Helena:

> So we made up our minds, intended to fight our way out or die in the attempt. Every face looked determined. We intend to burn the navy yard & shipping there … & blow down Gosport and Norfolk. The River was blocked & if we could not get out we sure'd burn the ship and go on board the *Yankee* of New York a little steamer that came to our assistance.[10]

* * *

One of the ironies of the catastrophe that befell the Gosport Navy Yard—a place surrounded by insurgent forces, threatened by the build-up of hostile troops, isolated by the blockage of its river access and weakened by the seizure of its arsenal—is that right up until its last day, a trickle of newly detailed officers arrived and reported for duty. Commander Livingston, the yard's new executive officer reported on 19 April, the Friday after Alden and Isherwood's efforts to retrieve *Merrimack* failed. And on that fateful Saturday, 20 April, Lieutenants Enoch Powell and John Irwin reported for duty. A witness to the surrender of the Pensacola Navy Yard, Irwin would again experience a navy yard's fall to Southern secessionists.[11] These officer movements were only possible because, despite the chaos spreading across Virginia, the Bay Line boats were still regularly steaming up and down the Chesapeake Bay.

The lieutenants' arrival at Gosport, while welcomed by its remaining officers and bringing a whiff of normality, couldn't mask the yard's approaching crisis. *Yankee*'s attempted seizure was the final stroke in the continuous drumbeat of threats deployed by the insurrectionists. To Commodore McCauley's overworked mind, the tipping point had finally been reached. First the channel blockages and fiasco of *Merrimack*'s attempted removal followed by the mass resignations of his Southern officers. Then the growing hostility toward visits to Norfolk, the constant rumble of troop trains, the seizure of the arsenal and the purported sighting of menacing earthworks at St. Helena. During the three-day period since Commodore Paulding's flying visit—it seemed like far longer—no new orders or instructions had been received from Washington. And the often-promised extra crews had never materialized. Without them, it was impossible to move to safety any of the ships ready for sea now lying at the wharf.

Commodore McCauley together with Commodore Pendergrast took stock of the rapidly changing situation at Gosport. Their prospects for holding off the insurgents did not look good. Up to this point they had done their best to place the yard in a defensive mode. The men under their command had removed what ordnance stores they could and destroyed the rest, and spiked more than a thousand of the yard's guns to prevent their immediate use by the enemy. When later testifying about the arrival in Norfolk of thousands of insurgent troops, as previously noted McCauley said "I was then satisfied that I had better get the ships out of their way, but it was my intention to remain and defend the yard."[12] The commodore's statement makes it clear he had no thoughts of abandoning or destroying his navy yard, but he now realized it was time for decisive action to get his vulnerable warships out of harm's way. Which ships were at risk and what threats did they face?

Of all the ships afloat in the Elizabeth River that Saturday, only a few of them were in immediate jeopardy. *Cumberland* was fully manned, guns at the ready and had a tug at hand if she was forced to flee under hostile fire. Of the five older frigates and liners in ordinary, all were unarmed, housed over, emptied of stores, completely unrigged and of little immediate military value to an enemy. The mighty *Pennsylvania* was stuck in the mud, but she was partially manned, mounted all of her 120 guns and given her tall sides, was relatively defensible. In any event, she couldn't be moved. So that left the core targets of the insurrectionists, the four warships lying along the wharf, unmanned but partially equipped and nearly ready for sea: the steam frigate *Merrimack*, the sloops-of-war *Germantown* and *Plymouth*, and the brig-of-war *Dolphin*. Of these vessels, *Merrimack* was wholly unarmed, *Germantown* and *Dolphin* fully armed and *Plymouth* carried no more than six of her guns.[13] For a provisional navy with no ships of its own, they were very tempting prizes.

The secessionist threats these vessels faced were two-fold. McCauley was convinced to the end that "there was no doubt ... they were erecting batteries on the St. Helena side."[14] He was only partially correct. They had not yet begun their work but the commodore's instincts were spot on—it was only a matter of time before the earthworks appeared. When they did, the men behind them would be aiming their guns at *Cumberland* and perhaps *Pennsylvania*, all to force the yard into submission. They would almost certainly avoid firing on vessels they coveted for their navy. The hazard to these four ready-for-sea warships that so worried McCauley was capture by boarding, a scenario which had played out that morning when the insurrectionists tried to seize *Yankee*. The threat of boarders was familiar to any navy man; it was an ancient and honorable way to conquer an enemy and his ship. This mode of combat would win the U.S. steamer *Harriet Lane* for Confederate forces in Galveston Bay on New Year's Day 1863. McCauley's handful of sailors and marines were far too few to mount other than a token defense against determined mobs of armed boarders. And if a ship were captured, she could be stolen or, given enough time, her guns turned against the yard's defenders. Now that most of Gosport's recently resigned officers were in the service of the secessionists, and many of them familiar with his vulnerable ships' capabilities, it was no idle worry. So what options did McCauley have to protect his vulnerable ships?

McCauley's best option—to man the ships with skeleton crews and sail them away—was no longer possible. Pendergrast best summed it up, "We had no pilots, no officers, no crews."[15] And even if all of the needed men had suddenly appeared at the yard's gates, any vessel attempting to flee would now be forced to run the gauntlet of hostile fire, insurgent boarding parties and more channel obstructions.

His extreme or worst-case option was to destroy the ships. That meant burning them, but the flames roaring up their tall masts would be impossible to confine to just a handful of ships. Putting the torch to highly flammable wooden warships would expose all nearby vessels to the threat of fire, and indeed, the entire navy yard and with it, all the surrounding workers' dwellings in Portsmouth. This dire option was only something a yard's commandant could justify if his navy yard was to be sacrificed—which is just what Commodore Paulding did in the wake of his abortive rescue mission. But McCauley hadn't yet reached that level of desperation.

McCauley's middle option was to scuttle his vulnerable ships. Scuttling a vessel of this era in the shallow, sheltered waters of the Elizabeth's southern branch was not as drastic or destructive as it first sounds. Bore a few holes in the vessel's bottom timbers and she will slowly and gently settle on the muddy river bed. Depending on her size and the location, the ship's upper deck would be just awash or still above water. Wooden ships were quite rugged and capable of withstanding such treatment. Plug the holes, pump them out and they were ready for service. And here it should be mentioned that Gosport Navy Yard had two diving bells in its inventory to facilitate raising a sunken hull.[16] An example of intentional scuttling that must have been known to the commodores was of Perry's two brigs which fought at the Battle of Lake Erie, *Lawrence* and *Niagara*. After the war, they were both scuttled in the harbor where they had been built to preserve their hulls.[17]

When queried about the difficulties of refloating scuttled vessels, Commodore Pendergrast answered: "They could, without very great expense, have been recovered and made effective."[18] McCauley was rather more pessimistic, answering a similar question but focused on quickly salvaging the ships: "It would have been attended with a great expenditure of time and money."[19] But even with the extra work necessary to recover scuttled ships, it was an option with important advantages to beleaguered defenders. Vessels so immobilized could not be stolen by an enemy. And with their guns spiked and magazines below the water, even if they were boarded and captured, they would be useless to the rebels as batteries. Given their perceived menace and that they were wholly unaware a relief force was on its way, Commodores McCauley and Pendergrast would have viewed scuttling as a rational option.

* * *

> At the time I did think it practicable to hold the yard, certainly until the next morning ... I had determined to remain there until attacked, and defend the yard with my small force to the best of my ability; and if the defense had proved unavailing, I would have retreated to the *Cumberland* and acted as circumstances might require.[20]

With these words, Commodore McCauley neatly summarized his defensive stance as of noon, Saturday 20 April. Orders for scuttling his most vulnerable warships had been issued to the yard's reluctant carpenters who would soon begin their unhappy work. At this moment, it is appropriate to pose two historical "what if" questions. First, consider how differently McCauley would have acted had Secretary Welles dispatched an officer the afternoon before to bring word of Paulding's rescue mission. Lieutenant John Irwin arrived Saturday morning, presumably on a Bay Line steamer about eight hours before the scuttling began. Why wasn't he or another well-briefed officer sent to throw a lifeline to the commandant of the Navy's most valuable dockyard? Second, reframing the question in the opposite direction, why didn't McCauley attempt to get word to federal forces at Fortress Monroe that he would soon be forced to scuttle his ships to avoid their seizure by the insurgents? *Yankee* could have reached the fortress in two hours. Neither of these speculative deeds occurred, but if they had, let your mind wander and imagine how much differently Gosport Navy Yard's history would have been written.

Five days after the scuttling, Commodore McCauley presented an official account of his actions at Gosport to Secretary Welles in which he succinctly described what happened next:

> I then commenced scuttling the *Germantown, Plymouth, Dolphin,* and *Merrimack,* destroying engine and machinery of the latter, cutting away the large shears, spiking all the guns in the yard and on board the ships in ordinary, including the *Pennsylvania,* and destroying such arms of the old and obsolete pattern as could not be placed on board the *Cumberland* and throwing them overboard, making the destruction of other things, with the exception of the public buildings, as complete as possible. [21]

Contemporary histories of the loss of Gosport tell that the first vessel to be scuttled was *Merrimack.* Sometime before noon *Cumberland*'s carpenter together with yard mechanics boarded the steam frigate.[22] The scuttling of the big ship was achieved by opening her sea valves—probably the bilge blow valves.[23] But by sinking her that way, as *Cumberland*'s Lieutenant Selfridge pointed out, the frigate's future salvage would be greatly simplified which turned out to be a prescient observation. How her engines were disabled is unknown, but it couldn't have been too destructive as her captors were able to restore them to service without undue difficulty. Even after she fully flooded and sank, her great draft ensured that *Merrimack*'s gun deck remained above water. [24]

At 12 o'clock the navy yard's bell was rung, the signal for its workmen to knock off for dinner. Once they left, the gates to the yard were locked and from then on, no one was allowed to enter without the commandant's approval. With Gosport's grounds effectively isolated from the outside, McCauley's men continued the work of scuttling with renewed vigor. Lying along the wharf under the yard's great sheer legs, *Germantown* was the next object of their attention. The accounts all say that her standing rigging was first cut away which is puzzling as it had nothing to do with flooding her hull. But there might be an underlying reason why this was done: the support guys of the sheers—a pair of huge, towering spars joined at their apex by a massive lifting tackle—were severed and the sheers collapsed on *Germantown*. The impact broke away two of her masts and shattered her upper deck bulwarks.[25]

This apparent act of gross vandalism seems starkly out of character with McCauley's intention to preserve his scuttled ships until it is grasped that the sheers were the yard's primary heavy lift system. They were indispensable for hoisting and mounting heavy naval guns.[26] With them disabled, should an enemy occupy the yard it would take many days if not weeks to repair and re-erect them and, in the meantime, none of the guns could be loaded out or easily removed.[27] The cutting away of *Germantown*'s rigging may have been an attempt to lessen the damage to her masts from the sweeping force of the sheers which, as it happened, didn't work out very well. But who had experience doing that sort of job?

The carpenters continued along the wharf and next performed their mischief on *Plymouth* and *Dolphin*, setting both their hulls on the river bottom. And although unmentioned by McCauley's or Paulding's reports, at some time that afternoon the two liners, *Columbus* and *Delaware*, were also scuttled. Lying in ordinary in the upper part of the yard near the graving dock entrance, it is not clear why these two old hulks were sunk. They were toothless and of little value to the secessionists. In any event once scuttled, because of their hulls' depth they did not fully submerge.[28]

* * *

Although locked out of the Gosport Navy Yard, the people living in its vicinity now sensed that something terrible was about to happen inside. Portsmouth resident and former navy yard employee William Peters wrote his recollections of his actions on that fateful Saturday afternoon:

> At about 2 o'clock P.M. on April 20, 1861, the late William H. Spooner came to my place of business in Portsmouth and said the Navy Yard gate had been closed and that none but those having special permits were allowed to enter the yard. He also said it was rumored that the authorities of the yard were making preparations to destroy that establishment with fire.

> To satisfy ourselves as to the truth of these rumors, Mr. Spooner and myself procured a boat and sailed in it up to and in front of the yard as far as the timber-dock. We saw enough to satisfy us of the truth of all we had heard. We found that the ships alongside the wharves had been scuttled and were slowly sinking. Men were engaged in cutting away the standing rigging of the sloop of war *Germantown*. Her upper masts were tottering. It was seen that a force of workmen were at work breaking off the trunnions of cannon.
>
> On our return in passing the brig-of-war *Dolphin* and receiving-ship *Pennsylvania*, we observed small guns on the poop-decks of those vessels, which, no doubt had been recently mounted there to repel some apprehended attack.
>
> Our people, already much excited, became alarmed at the situation of affairs at the Navy Yard, and the reported impending destruction of the naval establishment. It was feared that the burning of the Navy Yard would involve the destruction of Portsmouth.[29]

A citizen's committee was appointed to approach Commodore McCauley and persuade him to reconsider his intentions, but they were met with a locked gate and two of *Cumberland*'s Southern officers who had just resigned their commissions. The officers advised the committee that the commodore would refuse to meet with them. He was far too busy.[30]

By late in the afternoon McCauley's men had finished their work. The scene along Gosport Navy Yard's long wharf was unlike that any Portsmouth resident had ever before witnessed: a row of once proud men-of-war heeling at bizarre angles, their deserted decks barely above the water, the massive structure of the yard's sheer legs sprawled across *Germantown*'s broken hull. Barely a week had passed since Fort Sumter's bombardment and Virginia had approved its ordinance of secession just three days before. What woes had befallen their homeland? What horrors did the future hold?

Far removed from the chaos blighting the Elizabeth River and endowed with the gift of hindsight, Secretary Welles offered up his judgment through the medium of his diary on the defensive precautions taken at Gosport by Commodores McCauley and Pendergrast:

> McCauley and Pendergrast committed a lamentable mistake. They were deceived without doubt and in that terrible crisis were not equal to the emergency. They were not partisan politicians, and could not

> believe that so wanton, causeless, and extensive conspiracy existed; and when the crisis came, they were confounded and not prepared to act. When they did act, it was in bewilderment and error. Whether different officers would have had better success cannot be known. They might have rescued the *Merrimac* and some other vessels, though that is uncertain, for the Rebels had been long preparing for the event, and were the positive element; the Union men were passive. The Rebels were resolute and acted on the offensive; our officers were incredulous and on the defensive. They were anxious to strike and fight while others merely depreciated and repelled.[31]

Welles' after-the-fact views were somewhat self-serving but he was absolutely correct on one score: the secessionists acted decisively and with vigor to secure their homeland; the federal officers' responses, for better or worse, were always one step behind.

* * *

CHAPTER 17

The Die is Cast

Pawnee turned in a good passage steaming down the Chesapeake Bay, eventually rounding up off Fortress Monroe at 2:30 p.m. on Saturday, 20 April 1861.[1] But "good" is relative; a Bay Line steamer leaving Baltimore at the same time *Pawnee* sailed from Washington would have been alongside the fortress at about 6:30 a.m. Once again it boiled down to those eight hours—they would have made all the difference to Gosport and her harried defenders.

Commodore Paulding together with Army Captain Wright landed at the fortress and met with its commanding officer, Colonel Justin Dimick. The colonel was in no mood to give up any of his depleted garrison of regulars, but in accordance with General Scott's commitment he was willing to release Colonel David Wardrop's newly arrived 3rd Massachusetts Volunteer Regiment.[2] *Pawnee* cleared her decks and pulled alongside the wharf. To the wonderment of her old salts, a total of 349 men were promptly marched on board. All the while *Pawnee*'s lookouts searched for a steamer rumored to be bringing reinforcements from New York, but the horizon remained clear. By 6:45 p.m. she had cast off and was steaming toward Norfolk. Within minutes the sun dipped below the horizon, bringing on the fading light of an evening sky.[3]

Her open decks packed with undisciplined Massachusetts boys, her crew at their quarters with guns loaded, *Pawnee* began threading her way up the Elizabeth River. Although she carried no local pilot, the steam sloop was guided by her experienced master, Captain Mitchell. In a consummate display of navigational skill, he unerringly steered her toward the distant spires of Portsmouth and Norfolk, slowly working her way around the channel obstructions in the gathering twilight. So far Commodore Paulding's luck held. No opposition or disturbance greeted *Pawnee*, which was most fortunate as fighting such an overcrowded warship crammed with combustibles would have been fraught with horrific consequences.[4]

That troubled Saturday had brought many rumors to the harried defenders of Gosport. Most worrisome to the yard's guardship *Cumberland* was that of a steamer from Richmond said to be filled with armed men and headed their way. At 8 o'clock drums suddenly beat to quarters, alerting *Cumberland*'s crew to approaching danger. Lieutenant Selfridge raced to join his men at the forward 10-inch pivot gun. They gazed down the river and in the fading light could make out a large steamer standing toward them. Was she hostile? No recognition lights appeared in her rigging. Anxious hails bellowed from *Cumberland*'s deck. No answer. Young Selfridge sang out to his captain, "Shall I fire, Sir?" Another hail cried out. And then from across the river, came the welcome shout from the mighty *Pennsylvania*, "Do not fire, it is the *Pawnee*!"[5]

Cheers echoed across the Elizabeth, first from *Pennsylvania*'s overjoyed crew, then returned by the flagship, *Cumberland*. Moments later *Pawnee* nudged alongside the navy yard's wharf, her crew anxiously gazing at the drunkenly twisted row of sinking warships as they made the steamer fast. Commodore Paulding's expedition had narrowly avoided disaster, all because of *Pawnee*'s failure to show a signal of recognition when approaching a vessel that might have viewed her as hostile. Captain Wilkes, who witnessed the incident from *Pawnee*'s deck, later wrote, "It was only by the stentorian voice of our Commander … that we undeceived them as to our character."[6]

Within minutes Lieutenant Donaldson of *Pennsylvania*, the receiving ship's only remaining line officer, climbed aboard *Pawnee* and quickly explained to Paulding's incredulous officers that just hours before McCauley had ordered his ships scuttled "to prevent them from falling into the hands of the insurgents, who had for some time past been making demonstrations against … the navy yard." While Donaldson's words were being digested, Captain Wilkes together with Captain Wright jumped ashore to find Commodore McCauley and report Paulding's arrival. In the gathering darkness, the two officers began making their way toward the commandant's quarters, witnesses to mutilated guns, broken firearms and an atmosphere of disorder untypical of a navy yard. They soon ran into the commodore. Wilkes told him of Paulding's "arrival with assistance" and introduced Wright as the officer responsible for the yard's defense. Captain Wilkes recorded the details of their exchange in his official report:

> Commodore McCauley informed me that he had been deserted by all his officers, including an officer of the marines, and that the yard was without defense; that he had no one to rely on, and desired me to report that he had scuttled all the ships about 4 o'clock p.m. and

> had destroyed a large amount of property to prevent them from falling into the hands of the disaffected, led by the same officers who had left him.[7]

McCauley's remark about the marines is telling. Recall that after the wave of Southern officer resignations, only one marine officer remained at his post, Lieutenant Colonel Edelin, and as Wilkes later described, Edelin "had done no duty having been ill and a partial loss of Mind."[8] McCauley's marines were rudderless when their protection was most needed. Wilkes adopted a more scathing tone toward McCauley ten years later when he penned his autobiography. Charles Wilkes, always the "stormy petrel," was never afraid to speak his mind:

> I met him [McCauley] on my way to his quarters. He was armed like a brigand—swords and pistols in his belt and revolvers in his hand. I delivered the orders I had received and accompanied him to the *Pawnee*. His condition was such as unabled him to comprehend the situation and great difficulty in walking steady. He was unfitted for all duty.[9]

Wilkes' indelible image of McCauley is the one which was so often remembered and fits the circumstances then prevailing at Gosport. But the old commodore had not gotten a good night's sleep for days, was exhausted from single-handedly managing a besieged navy yard and in constant anxiety about the insurgents' threats. Captain Wright's account helps to round out the picture of Commodore McCauley, a man still game to defend his navy yard:

> On reporting to the Commodore of the yard I found him disposed to defend the yard and property to the last, and the troops were accordingly landed and some dispositions for defense taken.[10]

McCauley's later testimony made it clear that he knew he had been relieved of duty that Saturday evening and, given his physical and mental condition, was probably glad of it: "Commodore Paulding came down and superseded me in the command of all naval forces there." His fellow commodore, Garrett Pendergrast, snug on board his flagship, expressed a surprisingly distant attitude toward Paulding's arrival, especially considering he was the flag officer in command of the Home Squadron: "Commodore Paulding came there with instructions ... and took the whole matter into his hands. I did not participate in the transactions."[11]

Their nightmare was over but Commodore Paulding's was just beginning.

* * *

Three days had passed since Hiram Paulding sailed away from Norfolk, and in that brief time, everything had changed. Then he was a messenger with "no power to command," now he was the undisputed master of all he surveyed. But what greeted the commodore's eyes in the thickening darkness did not please him, not at all.

During *Pawnee*'s voyage down the Chesapeake, Paulding and his cohort of hastily assembled officers had agreed on a Gosport Navy Yard rescue plan:

> In carrying out the orders of the Department, it was my intention to have placed the vessels named [*Merrimack, Germantown, Plymouth* and *Dolphin*] in the channel to protect it from further obstructions, and at my convenience take them under the guns of Fortress Monroe, or send them to sea, as might be most expedient.[12]

Now surveying the slowly sinking ships lining the navy yard's wharf—the very same four vessels so central to his plan—a feeling of intense "amazement and chagrin" overwhelmed the commodore. He was a victim of an adage coined after his time that cautioned "no plan survives first contact with the enemy," and in his case, the enemy was Gosport's defenders' exaggerated fear of secessionist strength.[13]

But now Commodore Paulding urgently needed answers. Could the ships be saved? Picked officers together with gangs of men were hastily dispatched to examine each vessel and to report whether the sinking could be stopped. With the exception of the brig *Dolphin,* it was found impossible to halt the flooding. By throwing a block into *Merrimack*'s hold, Lieutenant Henry Wise discovered that the water had already risen above her orlop deck.[14] Unexplained by contemporary accounts is why these teams, which included *Cumberland*'s carpenter, couldn't shut the sea valves that he had opened only hours before to scuttle *Merrimack,* the most valuable of all the ships. But by then she was already settling onto the soft river bottom and would sink no further.

Adding to the pressure on Commodore Paulding was the secessionists' response to *Pawnee*'s sudden appearance at Gosport:

> Not long after my arrival at the navy yard, a flag of truce was sent to me by General Taliaferro, commanding the forces of Virginia in that vicinity, intimating that they had means of aggression, and stating that if I would desist from the destruction of the public property they would permit the *Cumberland* to go out of the harbor without being fired upon.[15]

The commodore delivered a sharp reply to Taliaferro: "Any act of violence on their part would devolve upon them the consequences." Local residents who caught wind of their exchange quickly assumed that Portsmouth and Norfolk would be bombarded. For them, Saturday's "reign of terror" was intensifying by the hour.[16] However, Paulding's stiff riposte to the general could not mask his gnawing worries about his diminishing ability to escape from Norfolk:

> The narrow parts of the channel had already been filled by obstructions placed there … the work of filling it up was still going on, and if the *Pawnee* and the *Cumberland*, with all the force then at the yard, had been left for its defense, the channel would in a very short time have been so far filled up that both those vessels would have been lost to the government, and in a short time the whole force must have been overpowered.[17]

Within a matter of minutes, Commodore Paulding realized his scheme to relieve Gosport could not be carried out. McCauley's scuttling had dealt it a fatal blow. And without warships to station along the reaches of the Elizabeth River and protect it from further obstruction, his force was now hostage to the insurgents' entrapment. Even worse, with each hour that passed fresh troops arriving from Richmond would surely strengthen their ability to seize the weakly guarded navy yard he now commanded. Paulding laid out his perceived dilemma:

> In view of the condition of things there were but two alternatives presented to my mind—the first to leave the navy yard and ships in the hands of people hostile to the Government, for it was apparent that the yard could not be held by our means of defense; or, using the power with which I was invested, destroy the public property of every description.[18]

But had the commodore correctly analyzed his situation?

* * *

At 9:00 p.m. on 20 April 1861, the entirety of the Gosport Navy Yard was under Paulding's command, its grounds fully controlled by federal forces, and the yard's gates were locked and guarded. Throughout the day and into the evening's twilight, no insurgent forces could be seen apart from "half a dozen men who were apparently lounging around the gate from curiosity," and no earthworks had been spotted at St. Helena from *Cumberland*'s masthead.[19] On the debit side, all but 200 of its heavy guns had been spiked, large quantities of firearms had been removed and older weapons destroyed or thrown "overboard." Three of its four warships nearly ready for sea were awash, sitting on the river bottom, and the yard's sheer legs had been toppled on to *Germantown*. And the arsenal at Fort Norfolk had been robbed of its gunpower.

But Paulding's reinforcements were all firmly in place. Fresh detachments of marines had relieved their weary counterparts who had been guarding the yard and *Pennsylvania*, making for a total of 180 leathernecks on the ground. A regiment of 350 Massachusetts volunteers was now protecting the navy yard, backed up by the nearly 180 sailors at Gosport and the 470 officers and men of the sloops *Cumberland* and *Pawnee*, the entire force bolstered by a dozen newly arrived naval officers.[20] Keeping watch on the Elizabeth were two powerful sloops-of-war mounting a combined total of twenty-eight guns, one of them steam-powered and the other assisted by the chartered tug *Yankee*. Thus, the navy yard was secure, protected by nearly 1,200 men-at-arms, in an active defensive posture and facing no visible hostile force. Four of its most prized ships were immobilized but capable of timely salvage and its trove of cannon was still in government hands. Further, the steamers *Pawnee* and *Yankee* gave the Navy the ability to aggressively patrol the Elizabeth's access channel to prevent its further obstruction.

The greatly feared secessionist forces outside Gosport's gates consisted of not much more than the twelve militia companies recently activated under the command of General Taliaferro, totaling less than 1,000 troops, still supported by only their handful of field guns. A battalion from Petersburg had recently arrived amongst Mahone's non-stop train shuttles but was posted behind Norfolk to guard against federal forces landing from Fortress Monroe. No significant earthworks had yet been dug and flooding on the James River delayed the dispatch of heavy guns. The newly established Virginia provisional navy possessed not a single man-of-war, only locally impressed watercraft. And as far as "means of aggression" to prevent the U.S. Navy's departure from Norfolk, a contemporary source says that "Taliaferro had no means to prevent them from leaving, and in fact was very anxious to have them go."[21]

Given the reality on the ground, several compelling questions assert themselves. What drove Commodore Paulding's perception that Gosport Navy Yard could not be defended and that he was in imminent danger of being overpowered? And what information did he possess and was it reliable? The following extracts from the commodore's testimony in reply to similar questions provide the answers:

> Other than through the officers at the yard … I had no knowledge of any military force. I had no opportunity of ascertaining what force might be there. I had no apprehensions whatever from the threat implied in the message I had received [from Taliaferro].
>
> I saw no batteries and had no knowledge of any. My information in regard to them was derived from Commodore Pendergrast.
>
> Men had been sent aloft to the masthead of *Cumberland* at sundown … and had discovered no batteries.
>
> Armed men were said to be in the vicinity, but I did not see any, nor did I witness any hostile demonstrations, except by flag of truce.[22]

In summary, Commodore Paulding sailed unmolested into the Gosport Navy Yard after sunset, and in the gathering darkness, he could see no hostile forces. His subsequent knowledge of the insurrectionists' strength was based solely on the flawed information fed to him by the yard's few loyal but anxious and exhausted officers. All of Paulding's first-hand sources of intelligence caused him no alarm. Although he soon came to the conclusion that destroying the navy yard was his best option, the voice of prudence should have cautioned that he had much to gain and little to lose by waiting a few more hours until daylight revealed Gosport's true situation. But the commodore thought otherwise.

Of all the officers present that evening at Gosport, none has written a more vivid narrative of what happened, both as found in the *Official Records* and his autobiography, than Captain Charles Wilkes. The seniormost of the dozen officers who accompanied Paulding on *Pawnee* and the one assigned to bring out *Merrimack*, Wilkes' prominence gives his observations added gravitas:

> There was no guard or any concourse of the rebels to oppose us any where. The *Cumberland* was a sufficient force to protect the whole and where to await day light to give an opportunity of seeing and ascertaining how matters stood, but it appeared that this was not to be.
>
> I was well disgusted with all this. … A kind of council of war was held, but there was nothing and no plan submitted. It was held in the dark and on the wharf. No satisfaction was to be seen in the faces of anyone, but we were informed that destruction was the order of the day.

> I was astonished as well as ashamed of the entire mode of doing business. We were all in the dark and no one knew the locality.[23]

Wilkes' sketches of the chaos surrounding Commodore Paulding's decision-making—or lack of it—made no mention of a potential option that the commodore never seriously considered. Recall that Paulding's original Gosport rescue plan envisaged removing four of the yard's warships and mooring them at intervals along the Elizabeth's vulnerable channel, both to prevent its further obstruction and to get the ships closer to their ultimate safe harbors. McCauley's scuttling derailed this scheme, but Paulding had another way of achieving its principal element, guarding the channel, by using the steam sloop *Pawnee* to actively patrol the river. Subsequent questioning of Paulding by Senator Grimes during the Senate committee's hearings touched on this possibility and their exchange revealed a startling admission by the commodore:

> Question: Is it your opinion that it would have been impracticable or impossible for *Pawnee* to have been successfully used in preventing obstructions from being placed in the channel by the enemies of the government?
>
> Answer: Without the power of making war upon the people so engaged it would have been impossible. …
>
> Question: You use the expression "the power of making war." Were you under instructions not to use force or would you have had any hesitation in using force to prevent the placing of obstructions in the channel so as to preclude your egress to the sea?
>
> Answer: I had no instructions in regard to that matter. I would have assumed the power to prevent it, but not to the extent of hazarding the lives of the individuals thus engaged.[24]

Secretary Welles' 18 April orders to Commodore Paulding squarely addressed this point: "you are invested with full powers to command the services of the entire naval force, and you will, if necessary, repel force by force in carrying out these instructions."[25] Paulding's orders explicitly authorized him to use force and yet under sworn testimony, he claimed otherwise. Why so? Could it be that Secretary Welles had verbally forbidden incurring any civilian deaths in the fulfillment of Paulding's orders? The Virginia watermen who sank hulks in the Elizabeth River were at best irregular militia, certainly closer to the status of civilians than bona fide combatants. Stepping back from the black and white of orders and sworn testimony, in these opening days of what would prove to be an incredibly sanguinary four years of civil war, the antagonists had yet to inflict significant bloodshed. The first U.S. Navy officer to die from hostile fire during the Civil War is widely acknowledged to be Commander James H. Ward, who was killed in action on 27 June 1861, well over a month into the future.[26] If Commodore Paulding was hobbled by such a restriction, real or imagined, in exercising his command authority, it might well explain why he failed to exploit the full naval potential he possessed. Without the imagination to grasp it or the will to repel force with force, he denied his relief expedition its best chance for success and doomed the Gosport Navy Yard.

* * *

The die had been cast. Gosport was to be burned. The commodore had decided:

> In a very short time I made up my mind that a necessity then existed for destroying the public property … It was a sad necessity but deeming it an imperative duty, I did not for a moment hesitate in its performance.[27]

Paulding took a scant few minutes to come to his determination, the most consequential of any he made during his long naval career. A rough timeline for his thinking is provided by testimony from Lieutenant Wise who, it will be recalled, was tasked soon after *Pawnee*'s arrival to assess *Merrimack*'s condition. Running to the big steam frigate and finding she had sunk, Wise immediately returned and reported his findings to the commodore: "He told me then it was time to mutilate the guns, and blow up the dry dock and fire the yard and ships."[28]

Although a "kind of council of war" had been held on the wharf, Commodore Paulding's decision was not embraced by all his officers. The outspoken Lieutenant Selfridge later recorded his misgivings:

> I thought at the time, and have had no reason to change my opinion, that the evacuation and partial destruction which followed was an irretrievable blunder … the Yard could have been held by the *Cumberland* and *Pawnee*, which had 100 Marines and a regiment of Massachusetts volunteers on board, until the ordnance stores and the *Merrimack* could have been removed or effectively destroyed. Meantime the channel could have been easily guarded

> against further blocking, by manning and employing some of the smaller vessels present.[29]

Also among the doubters was Paulding's senior officer, Captain Wilkes:

> I took the opportunity offered me of stating it was unnecessary destruction and in my opinion, everything of value, guns and all else, might have been saved and embarqued in vessels lying alongside the yard and thus avoid destruction. But a panic had seized the officer in command …
>
> I believe every other officer deplored this untoward act on the part of Paulding.
>
> I was fully disgusted with all that had taken place and well satisfied if any other officer had been in Command the result would have been different …[30]

They could think what they wished and say what they wanted, but it made no difference. The commodore's decision was final.

* * *

CHAPTER 18

Paulding's Destruction

If Commodore Paulding's order to put the torch to the Gosport Navy Yard could be criticized for its lack of forethought, the same could not be said of its subsequent execution. The need for incendiary materials and explosives had been clearly identified before *Pawnee* sailed from Washington. And that his officers were "detailed immediately to carry it into effect" only minutes after Paulding issued his orders gives the strong suggestion of prior consultation and planning. They may have been taking their drams in the dark and complaining of the chills, but once the word was passed, Paulding's officers coalesced into focused mission teams individually directed by Captain Wilkes who was given overall charge of the work.[1]

A group of 100 men dispatched by Commodore Pendergrast from *Cumberland* and joined by a party from *Pawnee* soon began the grueling job of breaking the trunnions off dozens of 9-inch Dahlgren guns. The ring of sledgehammers quickly filled the night air but the new guns' metal was so tough it resisted their hammer blows. Frustrated, they resorted to the less effective remedy of spiking. In partial consolation, the older 32-pounders' trunnions gave way to their sledges "like glass."[2]

Another team, this one led by Commander Rodgers and Captain Wright, was assigned to the most substantial of Paulding's targets for destruction, the yard's 341-foot-long granite-lined graving dock. Completed in 1834 at a cost of $1 million, this marvel of civil engineering allowed for the dry docking of the navy's largest vessels—the mighty *Pennsylvania* was shown on dock plan sections and elevations to demonstrate this was so—and would be of inestimable value to Virginia's secessionist naval forces. Destroying this dock was the job envisaged for the barrels of gunpowder that crowded *Pawnee*'s deck. To aid the effort, Colonel Wardrop detailed forty of his troops to carry them to the yard's upper end and into the graving dock.[3]

The significance of this team's assignment is highlighted by the quality of the officers who carried it out. Lieutenant Henry Wise, an experienced ordnance officer, was originally chosen by Paulding for this job but was replaced by three other senior officers, all of whom volunteered for the mission. Senior Navy Commander John Rodgers, the namesake son of the legendary War of 1812 hero, led the team and was ably supported by the Army's Captain Horatio Wright of the Engineer Corps, and Navy Chief Engineer James King. To these three volunteers must be added Lieutenant John Russell, an ordnance officer from the Washington Navy Yard, chosen by Rodgers to serve as an aide to keep communications open. Captain Wright's report yields an excellent eye witness account of how his team tackled the job of blowing up one of the nation's largest graving docks:

> The dock, which is a massive structure of granite masonry, has a pumping gallery running along the back of one of the side walls, entering from the level of the bottom near the entrance gate, and terminating, as is understood, in the pumping house near the farther end of the dock. Under the circumstances, for want of time for preparation and the darkness of night, this gallery offered the only means for establishment of a mine. Had the dock been full of water this advantage could not have been availed of, but we found in it a depth of only about 2 feet. We accordingly proceeded to construct in this gallery a platform of such materials as could be collected to a height above the surface of the water, and on this we placed the powder (2,000 pounds) which we had brought from the ship, established a train from the gallery to the outside, and connected it with four separate slow matches.[4]

Although the team's officers appear to have had little personal experience with the mechanics of Gosport's graving dock, their choice of a gallery entrance adjacent to the dock's gates was well calculated. Bursting the dock walls at the point where the gates were hinged would ensure the impossibility of closing it off for months if not years, rendering the dock useless. The amount of powder documented by Wright was little more than half of the total carried on board *Pawnee*. But it was fully capable, as later commentators confirm, of flinging the dock's stone blocks beyond the navy yard's walls and showering them down on homes in Portsmouth. One can only admire the courage and determination of the men who mined the graving dock, working in pitch darkness, wading through thigh-deep water and struggling to wedge in place heavy barrels of gunpowder. When their work was done, all extra personnel were sent back to *Pawnee* with the exception of John Reynolds, an ordinary seaman, who together with Rodgers and Wright, settled in waiting for the commodore's signal to light the fuse.[5]

* * *

While Rodgers and his men were packing the dock culvert with explosives, another team of destroyers was getting into its stride. Out on the calm, star-lit waters of the Elizabeth River, the redirected Lieutenant Wise ordered one of *Pawnee*'s boats manned and supervised its loading with turpentine tanks and cotton waste. Once this work was done, the lieutenant was summoned down to Commodore Paulding's cabin and told the names of the ships he was to burn. Assisting Lieutenant Wise in his heartless task were Lieutenants Ledyard Phelps, Charles McGary and George Morris.[6] Wise picked the biggest of the vessels on his list as the first to prepare for burning:

> I shoved off and came alongside the *Merrimack*; went on board with an officer and some of the boat's crew, got together combustible material, such as cordage, rope, ladders and gratings, hawsers, &c., and laid them in the form of the letter V, before her mainmast to one of the offshore gun-deck ports. On top of this were laid ropes of cotton waste saturated in turpentine. I then flooded the decks, fore and after and the beams likewise, with the same liquid, leaving the ends of the waste ropes outside the port.[7]

While Wise's team was busy on board *Merrimack*, his lookout reported that "a large crowd of men were about to make a rush and fire into the boat." But apparently they thought the better of it and the lieutenant was able to press on unmolested. He methodically worked his way down the line of warships tied up at the navy yard wharf, laying incendiary trains in each until he came to *Plymouth*. Being partly submerged, he could only spread combustibles across her forecastle. While his team was so engaged, his lookouts reported seeing a great many people and small boats along the St. Helena far shore. Apprehensive but unhindered, his party then struck out across the water toward the ships moored in the stream. Here they prepared the frigates *Raritan* and *Columbia* for burning but passed up the venerable *United States*, judging her so decayed as to not be worth the expenditure of turpentine. The receiving ship *Pennsylvania* was also left out. She was close enough to the other vessels that she would easily catch fire. And the distant hulks of *Delaware* and *Columbus*, partially sunk at the yard's upper end, were spared.

Lieutenant Wise completed his rounds by 1:30 a.m. To avoid becoming a target of the unknown people across the river, he kept his boat constantly moving from the shadow of one ship to the next while he awaited the commodore's signal to set fire.[8]

* * *

The remaining two parties of arsonists focused their attention on the most accessible and visible of Gosport Navy Yard's assets, its ship houses, store houses, workshops and buildings. Commander Benjamin Sands was detailed the responsibility of readying for destruction the two massive ship houses, one of which enclosed the frame of the unfinished 84-gun ship-of-the-line *New York*. Towering above the navy yard, each enclosing an inclined slipway, the ship houses' structures incorporated vast quantities of timber well cured by the heat of many Virginia summers—giant tinderboxes awaiting the match.

All the yard's other buildings were assigned to the attention of Commander Alden's team. Spread out across the navy yard in an irregular grid pattern, the variety of structures within the commander's purview was daunting. First were the shops and lofts catering to the yard's blacksmiths, joiners, riggers, boat builders, sailmakers, coopers and carpenters as well as the machine shops, boiler shops and foundries. Then came the various store houses, sheds and buildings containing naval stores and supplies, clothing and provisions, timber, masts and spars, cordage, cables and hemp, tanks and staves, and iron and copper. An armory, a sawmill with its engine house and coal shed, the machine and engine house enclosing the dry dock's pumps and the old pitch house filled out the list of the yard's significant industrial structures. Added to these were the main entrance building enclosing the gate, a porter's lodge, marine barracks, the commandant's office, houses providing quarters for the yard's officers and the stables.

For Alden, the biggest question was where to begin. The commander's attempts to exert coherence over his bluejackets' efforts were soon thwarted by demon rum, the ancient curse of military discipline. His men all too easily found the spirit room in the naval stores house and from there it went downhill. Many so filled themselves with liquor that "they were unable to keep up the work of destruction." But somehow the commander persevered and did what he could to prepare the buildings for burning.

By 1:45 a.m., both Sands and Alden were able to report that all was ready. They were then directed to send everyone back on board except a core of men necessary to ignite the incendiary material. As with the other teams, once reshuffled they bided their time, alert for the commodore's signal.[9]

* * *

With the receipt of Sands and Alden's reports, Captain Wilkes could now confirm to Commodore Paulding that all four destruction teams had finished their preparations. This signaled the rapid re-embarkation of the Massachusetts Volunteers and the groups of seamen filling out the shore parties. This exercise was eased by *Pawnee*'s being alongside the wharf, allowing the troops to march directly on board. Returning *Cumberland*'s men and supernumeraries, however, required boat work. In the midst of this shoreside turmoil, the yard's marine barracks caught fire by accident. A quick decision was made to let it burn; any effort to control it might have prematurely ignited the rest of the navy yard.[10]

The confused scene at the wharf took an unexpected turn when Commodore McCauley's son, Edward, suddenly ran up to Commodore Paulding. With tears streaming down his cheeks, the youngster desperately begged for the commodore's help to convince his father to leave his quarters. All appeals for him to do so had been rebuffed and Edward was acutely aware that the yard would soon be burned. One can imagine Paulding rolling his eyes at being forced to once again deal with McCauley's failings. Perhaps because Commander Alden was familiar with McCauley from his attempt to remove *Merrimack*, Paulding tapped Alden to handle him:[11]

> Commodore McCauley was fully under the impression we had gone there with our force to hold and protect the yard. With that comforting idea he had gone to rest with his brave boy at his side for the first time after many sleepless and anxious nights; but when I called on him after midnight to tell him that the time for his departure had come, that we were about to apply the torch to the yard, he was completely overcome with chagrin and mortification, and exceedingly adverse to leaving his post without first making some stand to defend the yard against any and all comers.[12]

Alden's gentle persuasions combined with soft reminders of his fate if he refused to yield finally convinced the stubborn McCauley to vacate his quarters and, together with his son, board *Cumberland*.

Close to this point in his narrative of the evacuation of Gosport, Captain Wilkes' memoirs state that Paulding "received a deputation from the authorities of Norfolk, headed by the Mayor, to request that Commodore Paulding would not destroy the navy yard." Wilkes had a habit of relating incidences out of order and blurring the details; it is possible he was referring to the earlier approaches by citizens to wait on McCauley or possibly Taliaferro's flag of truce mission.[13] Whatever the veracity of Wilkes' account, the hard fact was that Paulding had no intentions of accepting terms from any outside party. His mind was firmly made up. Gosport would burn.

Meanwhile, one of *Cumberland*'s boats was brought alongside for the final act, to bring off the shore parties' personnel once they applied the torch and lit the dock's powder train. Captain Wilkes was placed in command, aided by Lieutenant Russell, and instructions were issued to the teams clearly designating the embarkation point: in front of the lower ship house. Meanwhile, Lieutenant Wise, assisted by Lieutenant Phelps, hovered in the shadows, poised to fire the ships in the river.[14]

The time had come to ready *Pawnee* for taking *Cumberland* in tow as a prelude to moving Paulding's little fleet away from the narrow waters off the wharf. The tide now served and with the arson teams awaiting the commodore's signal, not a moment was to be lost. Their biggest job was to wind *Pawnee*—to shift her head to stern so that she could steam down the Elizabeth and toward Hampton Roads. Wilkes' account stressed that a "great delay was occasioned in winding *Pawnee*" and that "this proved a difficult undertaking." Managing this complex maneuver in close quarters required good ship handling skills, and must have suffered from being performed in the dark with *Pawnee*'s decks crowded by the Massachusetts Volunteers. The temporary absence of some of her men didn't help. But finally, by 3:30 a.m., she was ready to haul off the wharf.[15]

Meanwhile out in the stream, *Cumberland*'s officers and men had worked aplenty to get her underway from the position she had occupied for the past four days. Early that morning the steam tug *Yankee* had been brought along her starboard side and made fast. *Pawnee* then moved off the wharf, eased her stern in ahead of *Cumberland* and began the heavy labor of getting the sloop's towing hawser on board. Finally, at 4:15 a.m. the job was done and *Cumberland* slipped her moorings. Freely afloat now, the little flotilla straightened itself out and, under power from both *Pawnee* and *Yankee*, got underway and slowly headed down the Elizabeth River.[16]

The dilemma driving Paulding's movements was that to avoid self-immolation, it was imperative his fleet get well clear of the towering ship houses and the mass of *Pennsylvania* before setting them afire, but he also needed to keep station close enough to retrieve the parties that lit fuses and applied the torches: Rodgers at the graving dock, Sands and Alden in the yard, and Wise on the water. It demanded a fine balance of location and movement. And with daylight not until 5:27 a.m., it was necessarily performed in the dark, a less-than-ideal circumstance. Five minutes after getting underway, the commodore judged the moment had come. At 4:20

a.m. on Sunday 21 April 1861, a rocket flared high into the sky above *Pawnee*—the signal to apply the torch.[17]

* * *

The team leaders sprang into action. Deep in the shadows of the graving dock, the fuse to the powder kegs was lit—then the mad rush to escape. Inside the yard, Alden touched off building after building, and Sands' men set fire to the huge ship houses. Once done they raced to the wharf. Out on the Elizabeth, Lieutenant Wise again chose *Merrimack* as his starting point:

> I touched off *Merrimack*'s train, and in less than five seconds the flames burst out simultaneously from every part of her, like a flash of gunpower, so that I scarcely had time to get away from her.[18]

Pulling clear of the burning steam frigate, Wise and his boat crew followed their earlier route along the row of doomed ships, lighting off one train after the other. They then crossed the channel, expecting every minute that the dry dock would blow up, and lit off even more vessels. Behind the wharf they had just left, flames furiously raced through the yard, the night's darkness transformed by their incandescent bursts of heat. In a matter of minutes, Alden and Sands and their men had joined Captain Wilkes in his boat. But where was Rodgers? Growing increasingly anxious in the furnace-like heat, Wilkes' attention was rivetted on the gap where his team was expected to appear. The intensity of the voracious cyclone of flame was now overwhelming:

> The conflagration was rapid, in vast sheets of flames, and dense smoke, which enveloped us from the *Merrimack*, soon made it evident that it would be impossible for anyone to pass through it; yet they might have made the attempt to reach us. It was a painful anxiety to see every moment the chances of their escape diminishing. ... Then, and with great reluctance, I gave the order to shove off. As we emerged from the smoke we caught sight of Lieutenant Wise's boat, which appeared as if passing through the flames. At this time the masts and spars of the *Germantown* were on fire, and portions of her hull enveloped in the flames from *Merrimack*. I directed the boats to pull out, and was followed by Lieutenant Wise, the large flakes of fire falling around us. We had scarcely got beyond the ship *Pennsylvania*, which was the lowest vessel, when the flames from the lower ship house, under which we had been lying, reached her sides, and shortly after she was enveloped in flames.[19]

While Wilkes and Wise gamely endured gusts of flame and smoke, their men pulling for dear life, *Pawnee* silently led her charges down the still reaches of the Elizabeth. All hands had been sent to quarters, her magazine and shell room opened and her starboard battery manned. *Pawnee*'s marines lined her deck, under arms, staring quietly across the water as the sloop steamed past the dark, low shorelines of Norfolk and Portsmouth. Moving slowly to allow the two boats to catch-up and mindful of the need to maintain steerage way in the river, *Pawnee* steamed unmolested on her course toward the channel leading out to Hampton Roads. Back lit by the shimmering flames consuming Gosport, clusters of men could be seen at intervals along the left bank, their muskets' gleam confirming they bore arms.[20] Angry at the Navy's malicious destruction of their source of livelihood, the residents of Portsmouth could only pray the inferno would not devour their homes:

> The immense ship houses, with their millions of feet of timber, were seething volcanos of flames, and the huge ship *Pennsylvania* was a pyramid of fire, while the burning *Merrimack, Dolphin, Germantown, Plymouth, Raritan* and *Columbia* and the large store houses added to the conflagration and lighted up the heavens with a lurid glare that was seen for thirty miles. To add to the dangers of the night the dwelling houses on the north side of Lincoln street in Portsmouth caught fire, and the whole city was threatened with destruction, which was only averted by a change in the direction of the wind. Occasionally one of the guns of the *Pennsylvania*, which had been left loaded by her crew, would be discharged as it became hot enough from the fire to ignite the powder, but, fortunately, no one was hurt by them, and amid all of this crackling of flames, booming of guns and deluge of falling sparks, the cry arose that *Pawnee* was about to bombard the city.[21]

* * *

But Paulding's mind was fully occupied with escaping Norfolk, not lingering to exert more pain. The commodore's flotilla got as far as Craney Island before Wilkes' and Wise's boats finally rejoined it, nearly seven miles from the navy yard. Captain Wilkes, Commanders Sands and Alden, Lieutenants Wise and Russell and their boat crews must have been relieved that their ordeal was over. Less pleasant was reporting the failure to bring off Commander Rodgers, Captain Wright and Seaman Reynolds.[22] The hope was that they would soon find their way to safety. By now it was apparent to all that the graving dock's much-anticipated explosion had not been seen or heard, causing some speculation that the team's absence was somehow connected.

With increasing apprehension *Pawnee* approached the obstructions off Sewell's Point, a principal source of anxiety for Commodore Paulding. Daylight was upon them when they reached the narrowed channel which, following a hasty check, was found to be further obstructed since *Pawnee*'s passage the evening before. The decision was made to anchor *Cumberland* above the barrier and for *Pawnee* to slip her tow, pass around it and make haste to Fortress Monroe to debark the troops crowding her deck.[23] This was a sound move. All officers and crew were exhausted, the barrier needed further surveying before *Cumberland* could be forced through, and the sloop was reasonably safe in the channel, within sight of Fortress Monroe. Finally, if help was needed the trusty *Yankee* was close at hand.

Far in the distance smoke billowed high above the abandoned Gosport Navy Yard.

Pawnee closed the few miles to the looming fortress, discovering that another steamer was also standing into its wharf. She was *Keystone State* under the command of Lieutenant Woodhull, the relief vessel carrying extra men and marines dispatched by Captain Du Pont from Philadelphia, one of those that Paulding had so hoped to encounter the afternoon before.[24] Nonetheless, she had arrived at an opportune moment.

Pawnee was alongside Fortress Monroe's wharf at 8:00 a.m. and swiftly landed the Massachusetts Volunteers. Commodore Paulding also left her, presumably to report his actions at Gosport to Colonel Dimick and seek fresh news from Washington. A good six hours passed before the commodore shifted his flag to *Keystone State*, taking with him one of *Pawnee*'s lieutenants to help pilot his new flagship through the channel. Leaving *Pawnee* at the wharf, *Keystone State* got underway from Fortress Monroe Sunday afternoon at 3 p.m.

Soon she was in the channel and up to the barrier. Pendergrast, Marston and *Cumberland*'s officers had had plenty of time to further survey the obstructions and it was decided to send hawsers to both *Yankee* and *Keystone State* and bodily drag the big sloop through the sunken wrecks. The various accounts of eye witnesses—Paulding, McCauley and Selfridge—all agree that the process took "some" or "several" hours to accomplish. It was far from easy, even in benign sea conditions and with the pull of two powerful steamers. Selfridge made it clear—brute force did the job: "After several hours … the obstructions were crushed by the weight of the ship being dragged over them." *Cumberland* was apparently undamaged by her ordeal and by 8 p.m. she had been safely anchored in Hampton Roads.[25] For the first time in weeks, her men could rest easy.

* * *

Knowing that only with the pull of two powerful steam tugs did *Cumberland* force her way through the channel obstructions, and that it took her hours to do so, reel back the calendar four days and ponder *Merrimack*'s fate if she had attempted to breach this barrier. The steam frigate *Merrimack* was an altogether larger vessel than the sloop *Cumberland*, nearly half as long and of almost twice her tonnage.[26] *Cumberland*'s draft that afternoon, laden with ordnance stores taken from Gosport, was probably about 20 feet, and *Merrimack* would have drawn at least that amount of water, the frigate's greater full load draft balanced by her not being burdened by the weight of her armament, full stores, bunkers and outfit. And yes, she was a 973-horsepower steamer but her inefficient machinery was only capable of delivering sixty percent of that to her propeller, rather less in their "hastily patched up" state and even lesser still if she had been brought to a standstill like *Cumberland*.[27] It was at that moment—*Merrimack*'s being stopped by the obstructions—that her lack of a cohesive deck crew commanded by experienced officers would have put her at great risk. A skeleton crew under Commander Alden could not have lightened her, set kedge anchors to help heave her off, or repelled the secessionist boarders who surely would have been drawn to the sight of the stranded steam frigate. If *Merrimack* had attempted to pass through the channel barrier—absent a full crew, the help of tugs and a dollop of luck—Alden may well have been forced to burn her to prevent her capture. And curiously, as will be later clarified, for the U.S. Navy that might have been a more favorable outcome.

* * *

With the Home Squadron's flagship now safely moored in Hampton Roads, Commodore Paulding ordered *Keystone State* to rejoin *Pawnee* at Fortress Monroe with *Yankee* anchoring close by. For them, the ordeal of Gosport Navy Yard's destruction was over, or perhaps more accurately, its kinetic first phase was behind them. Inevitably an accounting for what happened at Gosport would follow, but for now, it was necessary to sort out the detritus of the recent days. During what remained of that long and stressful Sunday and into the wee hours of the night, no one got much rest.

All the marines brought off from the navy yard and its receiving ship together with their sole loyal officer, Lieutenant Colonel Edelin, were ordered on board *Keystone State* for passage back to Washington. Of Gosport's surplus seamen, twelve were ordered to join *Pawnee*, any deficiencies on *Cumberland* filled up as needed and the residue sent across to *Yankee*, provided her master was agreeable, for passage to the New York Navy Yard. Flag Officer Pendergrast was ordered to move *Cumberland* to mid-channel off Fortress Monroe and keep an eye out for suspicious armed vessels, remaining there "until the further pleasure of the Department is known."[28] And finally, Commodore McCauley together with his handful of officers who remained loyal to the end were welcomed to what quarters could be spared on Paulding's ships.

In all, Commodore Paulding signed five orders that Monday morning before the 7:30 a.m. departure of his flagship *Keystone State* and the steam sloop *Pawnee* for Washington.[29] There Commodore Paulding faced the unenviable duty of explaining to Secretary Welles—indeed, and to all of the United States—why he felt obliged to destroy and abandon the nation's premier navy yard.

* * *

CHAPTER 19

The Aftermath

No sooner had Portsmouth's volunteer militia entered the charred remains of the Gosport Navy Yard than Lieutenant Charles F. M. Spotswood strode up to its still-standing flagpole, raised the state's flag and formally took possession of the navy yard in the name of the Commonwealth of Virginia.[1] Spotswood had resigned his commission in the U.S. Navy only three days before. And it had only been four days since Virginia passed its ordinance of secession. What momentous changes a few days had wrought.

Before Spotswood could bask in the significance of his act, he received an alarming report that the yard's graving dock had been mined and its ignition train spotted. The quick-thinking lieutenant immediately ordered the wicket gate open, flooding the dock and causing the barrels of gunpower that had been so arduously packed inside it to float out. Thus arose the tale that the secessionists had discovered and disrupted the powder train in time to save the dock. What actually preserved the dock was the compassion of a U.S. Navy sailor for his friends living in Portsmouth. The night before he had been ordered to light the fuse—which he did so he could claim he really had lit it—but instead of placing the sputtering fuse where it would ignite the train, he threw it into the water.[2]

As Gosport Navy Yard's new owners made their rounds, they were surprised to find how little damaged the yard's buildings and facilities really were. Lieutenant George Sinclair, one of Gosport's former ordnance officers, wrote of his first impression: "The most abominable vandalism at the yard. Destruction less than might be expected."[3] Later testimony by lifelong Portsmouth resident James Brownley supplied more details: "The marine barracks, the two ship-houses, the riggers' loft, the sail loft, and main entrance to the gate are all that were destroyed."[4] The evidence was that Commander Sands' arson teams were successful but that Alden's failed miserably. Later tallies by newly commissioned Paymaster Peters revealed that provisions worth $38,763 had survived the holocaust as did $56,269 of clothing and small stores and an astonishing $1,448,223 of general naval stores and supplies. Included in this latter haul were "large quantities of timber ... copper in sheets and in bolts ... iron in sheets and bars ... and a vast amount of miscellaneous articles in great variety and of inestimable value."[5]

The scene out on the still waters of the Elizabeth River registered a harsher reality. Ten U.S. Navy warships had been wholly or partially destroyed or abandoned on the morning of 21 April, the number rising to eleven if the fully framed liner *New York* inside Ship House A was counted (see Appendix 4). Of all the "Vessels of War of the United States Navy" listed in the 1861 *Navy Register,* the eleven ships lost at Gosport comprised twelve percent of the total. But if graded by the number of guns mounted, these eleven vessels cost the Navy twenty-six percent of its total firepower afloat. Portsmouth resident James Brownley's Senate committee testimony later sketched a riveting image of each ship's fate:

> The *Merrimack* after being scuttled and sinking as far as she would, was burnt to the water's edge; she has been cut down, I know, six feet below the copper line. The *Pennsylvania* was burnt to the water's edge, leaving about 23 ft, which was her draught, below water. The brig *Dolphin* drifted ashore and was burned down to about six feet. The *Plymouth* was merely scuttled and not injured by fire, and has since been raised by the insurgents; she is now dismantled and lying against the wharf. The *Germantown* was a good deal injured from the falling of the shears; two of her masts were broken away; she took fire from the ship-house, and was very much damaged, so much injured that they have taken off her guns and laid her out in the stream, and are unable to make any use of her. The *Columbus* and *Delaware* were simply scuttled, and otherwise uninjured. The *Columbia* was burnt to the water's edge, and is out of sight. The *Raritan* was also burnt to the water. The *United States* was not injured at all, and is now lying before the yard manned with fifteen 32-pounders; her guns bear directly on Portsmouth.[6]

Historians have minimized the destruction of so many of the Navy's ships, employing the arguments that only *Merrimack* had any real value, and that *Germantown, Plymouth* and *Dolphin* were obsolete sailing vessels. The remaining seven ships were simply dismissed as worthless old hulks rotting in ordinary. But such sweeping generalizations ignore the latent value within those old but stout oak-framed hulls. Navy Secretary

Toucey's final annual report, presented five months before Gosport's loss, recommended that eight of the Navy's liners be converted into steam frigates and that five of its sailing frigates be razeed into efficient sloops-of-war, on the model of *Cumberland*.[7] Such large cruiser conversions might have proved of limited value in furtherance of the brown water naval campaigns of the Civil War that followed, but in April 1861 no one could see that far into the future. And if the Union had become embroiled in a war with a European power, as seemed likely in late 1861, cruisers built from the hulls of *Raritan, Columbia, Delaware* and *Columbus* would have been valuable additions to a navy stretched to its limits (see Appendix 6).

What must have made the secessionists' hearts beat faster was the sight greeting them at the navy yard's gun parks: row after row of heavy naval guns, virtually all of them serviceable. From 11-inch guns down to the ubiquitous 32-pounders, some 1,200 guns were counted including nearly 200 of the latest Dahlgrens. And they were, as Lieutenant Sinclair crowed, "ready for our cause." Hasty efforts by the retreating federals to spike and maim the guns proved largely futile. The spiked guns were quickly returned to service and if Sinclair's count was correct, only eight of the 32-pounders had been destroyed.

Adding to the cannon clogging the yard's grounds, the marines' howitzers were fished out of the river together with substantial numbers of small arms and 2,200 of *Merrimack*'s 10-pound powder cartridges in watertight tanks. More than a hundred gun carriages were saved at St. Helena and "many thousands of shells and shot" of all calibers were found throughout the yard.[8] And recall that the navy's arsenal at Fort Norfolk had been cleaned out only days before, netting the insurgents all the gunpowder they needed.

Unlike the scuttled and burned warships in the river, which would take many months to salvage and restore to service, or the yard's workshops, which first needed refurbishing and then even more time before matériel could be manufactured in quantity, Gosport's guns were of immediate use. Since they were transportable, their value extended far beyond Norfolk and the Tidewater to all of Virginia, and when Virginia joined the Confederacy, to all of the Southern states (see Appendix 5). For the agrarian South with little heavy industry of its own, Gosport's haul of naval guns was a godsend. Admiral David Porter, who encountered the purloined guns everywhere he fought during four years of war, stated the bald truth:

> Had it not been for the guns captured at Norfolk and Pensacola, the Confederates would have found it a difficult matter to arm their fortifications for at least a year after the breaking out of hostilities

> ... Great as was, therefore, the loss of our ships, it was much less than the loss of our guns.[9]

During the later Senate committee hearings on the destruction of the Gosport Navy Yard, the Navy Department issued a summary statement itemizing the value of its lost public property. Although it is difficult to equate the tabulated amounts with today's prices, as a measure of the national wealth sacrificed by Commodore Paulding's destruction and abandonment of the navy yard, the total was equal to about seventy-five percent of the naval appropriations for 1861.[10]

TABLE 19.1 Summary Statement of the Public Property of every description at the Norfolk Navy Yard at the time of its Destruction.[11]

Expenditures for improvements and repairs at navy yard to 1 January 1861	$4,508,439.95
Expenditures for improvements and repairs at hospital	393,174.70
Expenditures for improvements and repairs at magazine and ordnance works	204,562.53
Estimated value of eleven vessels destroyed	1,980,000.00
Ordnance and ordnance stores	664,883.78
Provisions	55,148.42
Clothing	71,678.75
Small stores	8,023.14
Surgeon's necessaries, &c.	7,299.89
Miscellaneous articles	1,865,443.72
Clothing of marine corps	1,527.05
TOTAL	$9,760,181.93

* * *

The public reaction to the loss of the Gosport Navy Yard was more muted than might have been expected considering the immensity of the disaster. Two principal reasons account for this phenomenon. First, it was one of many severe crises simultaneously besetting the Union at that particular moment in time. Scanning the front page of *The New York Times* for 24 April 1861, the first day the news from Norfolk was comprehensively reported, it can be seen that the story of Gosport's burning competed with headlines of similar magnitude under "The War News," all indicators of the nation's dire circumstances:

- Virginia Troops Threatening Washington, the Capital Considered Safe.
- Martial Law Declared in Baltimore, the Business of the City Entirely Ruined.
- The Burning of Gosport-Navy Yard, Immense Amount of Government Property Destroyed.
- Departure of Over Three Thousand Troops from New York.

Somewhat trivializing *The New York Times*' reporting of Gosport's burning was its outsized emphasis on the participation of the steam tug *Yankee*, which had just returned to the city. The paper cast the New York registered tug in the role of "local boy makes good" as encapsulated by its sub-headline of "Tug Yankee Tows the Cumberland to Sea."[12]

Further blunting the impact of what happened in Norfolk was the unhappy fact that it represented the latest installment in a string of federal disasters to befall the Lincoln administration. Over the past ten days, the public had been confronted with sensational stories about the surrender of Fort Sumter, the calling up of 75,000 militia, Virginia's secession, the burning and evacuation of Harpers Ferry, the blockade of Southern ports, rioting in Baltimore and now the destruction and abandonment of the Gosport Navy Yard. It was but one more catastrophe to traumatize a numbed nation.

A second reason the debacle in Norfolk was received with so little emotion or outrage was that while its days-long drama culminated in a spectacular inferno and the dramatic escape of *Cumberland* and *Pawnee*, not a single life was lost. No effusion of blood or human sacrifice occurred, no uplifting heroic action ennobled any of the participants, it was all about a calculated destruction of "public property" to deny military resources to the Southern rebellion. Unlike the attack on Pearl Harbor eighty years later to which it has been compared, it signally failed to shock the average citizen or alter the nation's appetite for war.

Although the events at the Gosport Navy Yard didn't generate an immediate public reaction, they exerted a profound impact on the Union and the Civil War that followed. Most basic was the documented loss of nearly ten million dollars' worth of dockyard facilities, warships, naval ordnance, and material, all replaceable but all invaluable to an adversary lacking in manufactured resources, especially the guns.

At a higher level, the loss of a strategically located navy yard and port facility complex brought serious consequences. American forces struggled valiantly to keep Norfolk out of British hands during the War of 1812—recall Midshipman McCauley and the Battle of Craney Island—but failed at the first whisper of rebellion to maintain federal control of Virginia's most important seaport. The impact of its forfeiture would be felt by the Navy's blockading squadrons and the overworked naval dockyards that sustained them. Confederate threats to Hampton Roads would also impinge upon the logistics of the Union's future Peninsula campaign. And the ironclad spawn of Southern genius would on those same waters stun the nation with an awesome power wholly unimaginable in April 1861.

But perhaps most intangible was the loss of face, of prestige, of an aura of competence suffered by both the Lincoln administration and the Navy. Start with the political miscalls guiding the federal government's handling of Virginia's secession, its uncoordinated reactions to the crises of Forts Sumter and Pickens and its failure to timely recognize the vulnerable position of its finest navy yard at Gosport. Add to them McCauley's miscalculations in the face of a determined but greatly inferior foe and Paulding's failures, first to grasp the yard's precarious position and second in over reacting, destroying a navy yard and everything in it before the light of day brought clarity. Taken as a whole the image created by these miscalls, at all levels, was that of colossal blundering.

Other historians' evaluations mirror this same assessment. Charles Paullin's *History of Naval Administration* concluded that "the evacuation of Norfolk and the Norfolk Navy Yard was one of the greatest blunders of the war."[13] And Gideon Welles' biographer and contributor of the section about him in Coletta's *American Secretaries of the Navy,* John Niven, expanded on this theme, encompassing two of the most responsible parties in his conclusion:

> The blunders, uncertainties, and feeble policies of the administration that marked the Sumter-Pickens affair also characterized the attempted destruction and the abandonment of the Norfolk Navy Yard. ... Welles' orders were ambiguous, and he erred in his selection of the old and irresolute Captain Hiram Paulding to head the relief expedition. ... The loss of Norfolk was a disaster that instantly exposed the Department to political and editorial attack while initially weakening the Navy's capacity to perform its missions.[14]

* * *

All bureaucracies are creatures sustained by their consumption of documents and in this respect, the U.S. Navy Department of 1861 was no different. Officers in leadership positions who participated in the affair at the Gosport Navy Yard were obligated to submit written reports to their superiors. What stands out to the modern peruser of these reports is how promptly they were submitted, some obviously being composed while the author was still afloat, and the quality of their narratives, in terms of grammar, composition and flow. They all make good reading.

The first of the significant reports submitted was that of Captain Charles Wilkes, Paulding's second in command, whose report was dated 22 April 1861 and addressed to Flag Officer Paulding.[15] Written while he was still on board *Pawnee*, it covers three pages of the *Official Records* and is typical of Wilkes' descriptive and all-encompassing style—he rarely misses a detail nor shies from the prolixity to relate them. As it was an official document submitted to his superior officer, Wilkes didn't breathe a word of his criticism of Paulding's actions; that came later in his autobiography.

Next to document his actions was Commodore Paulding, whose report was dated 23 April 1861, but was submitted to Secretary Welles under a separate cover the next day.[16] In his report he signed his name over his temporary title, "Comdg. U. S. Naval Forces in the Waters of Virginia, pro tem," and in his submittal cover, he reverted to his rank of captain. To his two-page report must be appended a later document dated 2 May 1861, titled "Narrative of the destruction of the Norfolk Navy Yard, by Commodore Hiram Paulding."[17] As this account opened by stating Paulding has "not seen the editorial in your paper" it is clear that it was in response to a newspaper for the purpose of correcting "a misapprehension of the facts." Paulding's report gives an honest assessment of his expedition's actions and he presented himself as being absolutely convinced of the correctness of his decisions. His report is valuable, in part, because of its appended "List of Officers attached to and found on duty at the navy yard, Norfolk, Va., and on board the receiving ship *Pennsylvania* on the night of April 20, 1861." This short list brings home how utterly alone Commodore McCauley was when Paulding arrived—only three of his original line officers—Edelin, Donaldson and Semmes—remained by his side.

The third officer to submit his report was Commodore McCauley, but before he could do so a telling event occurred: on 24 April 1861 Secretary Welles issued a General Naval Order, which covered the findings of Commodore James Armstrong's court-martial stemming from his surrender of the Pensacola Navy Yard.[18] That Welles chose to issue his general order just days after the destruction of Gosport—before the reports of its participants had been received and the full ramifications of its loss could be digested—suggests his motivations were overtly political. And that Armstrong was found guilty of the charges laid against him, harshly sentenced including a loss of pay and subject to a reprimand served to amplify Welles' message to his audience—I am performing *my* duty, and I expect officers to remain true to their flag no matter what. Curiously, Secretary Welles inserted a Delphic statement near the end of his general order, alluding to what had just occurred at Norfolk. Referring to Armstrong's failures, he wrote:

> Had he discharged his duty with stern fidelity he might have disappointed some, but, as events have recently shown in the case of another officer placed in similar circumstances, he would have been upheld by the applause of his fellow-countrymen and the respect even of antagonists.

Which officer was Welles referring to, Commodore McCauley or Commodore Paulding? Or did he purposely intend to be vague?

Back to McCauley's report.[19] On 25 April the commodore submitted a workmanlike accounting to Secretary Welles that did not attempt to cover up his decisions to scuttle his vulnerable ships, cut away the sheers and spike his guns. But careful analysis hints that McCauley's choice of words might have been influenced by Armstrong's court-martial specifications, in particular his emphasis on not having the means to remove the ships, on being faced by overwhelming insurgent forces and on his purposeful destruction of public property rather than letting it fall into enemy hands. How well his report was received by the secretary would only be revealed with the passage of time.

The two other reports of merit include Army Captain Wright's account dated 26 April, with its superb detail describing how he mined Gosport's graving dock.[20] Unexpected is the chronicles of his odyssey after falling into the hands of his Virginia captors and the gentlemanly conduct observed by both sides in the very first days of America's bloodiest war. The final significant report was submitted by Commander Alden on 30 November 1861.[21] Alden evidently didn't believe he was obliged to write a report, but Secretary Welles thought otherwise and ordered him to do so. The result was probably the most compassionate of all the reports, giving a window into Alden's character. His sketch exposing the influence of "masked traitors" on McCauley is priceless and, after the lapse of time, Alden allowed that perhaps the commodore had good reason for doubting his power to safely remove *Merrimack*. *Tempus sanat omnia.*

* * *

Anyone familiar with the basic contours of Civil War history knows how this story ends. By 7 May Virginia had been admitted to the Confederacy. That new linkage energized the rapid conversion of the burned-out hulk of *Merrimack* into a fearsome ironclad ram which, in March 1862, challenged the U.S. Navy's supremacy at the Battle of Hampton Roads. How the Confederates arrived at their David and Goliath moment was directly linked to the Union's loss of the Gosport Navy Yard in April 1861. That calamitous event gave the South the two vital components needed to create an advanced warship—*Merrimack*'s intact hull and machinery and a fully functioning graving dock. Add to them the facilities of a first-class navy yard brimming with naval supplies, timber and ordnance, and the stage was set. Only the catalyst was now required: a visionary who had the imagination to grasp that for a new-born nation's navy to prevail against a more powerful, conventionally armed foe, a next-generation warship—a steamer clad in armor plate—offered its only hope of survival. That visionary was a former senator from Florida, now the Confederate States' secretary of the navy, Stephen R. Mallory, who less than three weeks after Gosport's burning informed his government: "I regard the possession of an iron-armored ship as a matter of first necessity."[22] Thus was planted the seed that produced the first ironclad warship to engage another ironclad in combat.

Merrimack's gestation as a "ship of destiny" was neither easy nor trouble-free. After being scuttled by McCauley she settled on the river bottom and so was partly submerged when Paulding applied the torch. Only her upper decks were destroyed; the frigate's stout oak-framed hull and its valuable machinery survived the inferno. Weeks later her wreck was raised and on 30 May her "burned and blackened hulk" was warped into Gosport's graving dock.[23]

Secretary Mallory had already hedged his bet on getting an ironclad. Two weeks earlier he dispatched Lieutenant James North to Europe with orders to buy an ironclad frigate from France.[24] And by early June he had estimates in his hands for what it would cost to restore *Merrimack* to her full glory as a steam frigate or to convert her into an ironclad. The difference, $450,000 versus $172,523, made his next step easier.[25] A frigate, even one as powerful as *Merrimack*, could be blockaded in the Elizabeth River but an ironclad would be the master of all. By late June his hand-picked team of a naval officer, a constructor and an engineer had come up with a design for a casemate ironclad that could be built by the South. It was but a small step to translate it, with modifications, to *Merrimack*'s charred hull and rusty engines. The savings in time would be a fair trade for dealing with her less-than-desirable deep draft.[26]

Secretary Mallory unhesitatingly authorized the work to begin on the Confederacy's first ironclad warship. In a whirlwind of effort, his enthusiastic team initiated gun-against-armor trials, sketched up the needed conversion drawings and began the labor of restoring *Merrimack*'s engines. Richmond's Tredegar Iron Works was stressed to its full capacity rolling out 2-inch-thick sections of armor plating, all punched to measure, totaling some 723 tons.[27] Five days after the last plate was delivered, the newly christened C.S.S. *Virginia* was floated out of Gosport's graving dock. The date was 17 February 1862.[28] Far from ready for battle much less finished, the gleaming grey mass of the South's latest warship signaled a new, brutalist approach to fighting at sea. Her sloped casemate was pierced for just ten guns—but they were the heaviest that could be found at Gosport—and topped by only a tall smoke pipe: no masts, sails or rigging cluttered this engine of war. She could never survive on the high seas but that made no difference—it would be good enough if she could sweep the Yankee squadron from Hampton Roads.

In late February Secretary Mallory appointed Captain Franklin Buchanan to command the James River defenses. In vesting Buchanan with such responsibilities—he would also captain *Virginia*, his flagship—Secretary Mallory put no gloss on his new ship's capabilities: "*Virginia* is a novelty in naval construction, is untried, and her powers are unknown ..." He then indulged in the boast that "her powers as a ram are regarded as very formidable." Mallory concluded his orders on a note of urgency: "Action—prompt and successful action—now would be of serious importance to our cause ..."[29]

Buchanan quickly got to grips with his new command responsibilities, selecting officers and appraising *Virginia*'s capabilities. He also scouted out his enemy and could plainly see that the Union squadron occupying Hampton Roads was strong. Two steam frigates *Minnesota* and *Roanoke*, sisters of his flagship in her earlier life, and the sailing frigate *St. Lawrence* held the anchorage off Hampton Creek, while *Congress*, also a sailing frigate, and *Cumberland*—the sloop that guarded Gosport before it was burned—watched the more distant waters at the mouth of the James off Newport News. Of all these ships he feared *Cumberland* the most. Flag Officer Buchanan had heard that only she was armed with the new 70-pounder Dahlgren rifled gun, the weapon with the best chance of penetrating *Virginia*'s 4-inch-thick armor plate.[30] Before the war, Buchanan had served as the Washington Navy Yard's commandant and knew all about Dahlgren's creations. And he had surely heard the stories of *Cumberland*'s threats to bombard Norfolk and how she alone had escaped the yard's inferno. It was an easy decision for "Old Buck" to make—*Cumberland* would be *Virginia*'s first target.

When Flag Officer Buchanan ordered *Virginia*'s officers to take her out of the navy yard and steam for Hampton Roads on the morning of 8 March 1861, only her executive officer, Lieutenant Jones, knew of Buchanan's intention to attack the enemy; everyone else thought it would be her first trial run.[31] Just as surprised were the U.S. Navy's five warships, idly at anchor that Saturday morning, some with laundry fluttering in their rigging. Less than a year after the debacle at Norfolk, the Navy would again be caught off guard by its Southern foe.

Although *Virginia* sortied before noon, her slow progress down the Elizabeth River ensured that it wasn't until nearly 1 p.m. before she was identified by Union lookouts. Out she came into the expanse of Hampton Roads but then veered to the west, leaving astern the line of frigates off Fortress Monroe. Looking more like a barn on fire than a man-of-war, *Virginia* inexorably pointed her iron-tipped bow toward *Congress* and *Cumberland*. After firing a first ranging shot at *Cumberland*, *Virginia* ripped *Congress* with her guns but not for long. The frigate's tormenter kept steady on her course for the true target, *Cumberland*.[32]

Cumberland's officers and crew frantically prepared for action. Her officer of the deck was the same energetic Lieutenant Thomas Selfridge who had served on *Cumberland* throughout her ordeal at Norfolk. The sloop's men sprang to her guns while off her starboard bow, the iron leviathan maneuvered into a raking position, pouring in fire. Soon irked by wasting precious powder on time-consuming cannonades, Buchanan passed the order: they would ram her! *Virginia* smashed into *Cumberland* just below her fore chains, driving a huge hole in her side. Caught hard by *Cumberland*'s heel, *Virginia* slewed alongside the doomed sloop, tearing off her ram as she swung free. Oblivious to their peril, *Cumberland*'s well-trained crew blasted three rolling broadsides into the ironclad, damaging her armor, shredding her funnel and topsides, smashing the muzzles of two guns, killing one of her crew and wounding others. But their bravery could not save *Cumberland*. Refusing calls for surrender, her captain cried back "Never! We will sink with our colors flying!" She soon plunged to the bottom, her flags still waving defiantly, a third of her men casualties.[33]

Virginia licked her wounds, laboriously turned around in the James and retraced her course. She closed in on the now-beached *Congress* and, virtually unopposed, sent shell after shell into the hapless frigate which reluctantly surrendered. As his men attempted to board *Congress*, Flag Officer Buchanan was shot in the groin, forcing him to hand over command to Lieutenant Catesby ap Roger Jones. Thoughts of finishing off the nearby *Minnesota* were briefly mulled but prodded by the pilots' fear of the growing darkness, her officers agreed that *Virginia* had achieved quite enough for one day. Her crew was exhausted—they could destroy the stranded steam frigate in the morning.

Lieutenant Selfridge later pled the case that by damaging *Virginia* and delaying her progress that crucial first day, *Cumberland*'s spirited action prevented even more carnage:

> I am firmly convinced that the great sacrifice made for the honor of the flag by the crew of *Cumberland*, though it failed to save their ship, resulted in the much greater achievement of saving the fleet if not the Union.[34]

After a sleepless night preparing for dawn's combat, *Virginia* was greeted by John Ericsson's turreted ironclad, *Monitor*. The two armored warships, one a giant and the other a raft mounting a pair of guns, fought an inconclusive duel lasting till noon. If the first day of the Battle of Hampton Roads had handed the U.S. Navy its most humiliating defeat since the War of 1812, the second day of battle brought deliverance.

The Navy's severe losses that first day—two of its finest capital ships and nearly 280 men killed or wounded—were inflicted by the ironclad *Virginia*, a novel engine of war birthed from the charred hulk of *Merrimack*, herself a casualty of the abandonment of the Gosport Navy Yard. The birds from that disaster had finally come home to roost.[35] But now pause to consider some imponderables of history: *if Merrimack* had either escaped from Gosport or been burned in the attempt, or *if* her hull had been totally destroyed by Paulding's men, or *if* the yard's graving dock had actually been blown up, the events that followed would have turned out very differently. And *if* that had been the case—no resurrected *Merrimack* to wreak havoc—future historians might well have consigned the loss of the Gosport Navy Yard to just one more Civil War footnote.

* * *

CHAPTER 20

The Accounting

Commodore Paulding's weary officers debarked from *Pawnee* and *Keystone State* at the Washington Navy Yard on the afternoon of 23 April 1861, bringing closure to their active participation in his abortive expedition to save the Gosport Navy Yard. If they hadn't already been written, post-action reports needed to be composed and submitted, interviews with Secretary Welles could be expected, and fresh orders anticipated. In the closing sentence of his official report, Paulding put his spin on how he expected those who returned with him should be treated:

> The officers who served under my command in the performance of the duty devolved upon me have a claim upon the respectful consideration of the Government for the intelligence, energy, and patriotism with which they gave me their support and assistance.[1]

But collectively these same officers had only two days before put the torch to ten million dollars' worth of public property and willfully abandoned the Navy's finest dockyard. Hovering in the minds of some was the unpleasant specter of questions asked about what they had done and why, or possibly a court of inquiry. They all knew the loss of the Pensacola Navy Yard had initiated that same sequence and it led to Commodore Armstrong's court-martial. And they would read, perhaps with some anxiety, Welles' general order posted the day after their return, confirming Armstrong's punishment and adding a reprimand to complete his public shaming. The three commodores, especially McCauley and Paulding, all had varying degrees of responsibility for what had happened at Gosport. Could they be facing the same ordeal?

But nothing happened.

No longer a commodore in title, Captain Paulding again took up his duties in Washington at the Office of Detail; Flag Officer Pendergrast, still in command of the Home Squadron, kept busy chasing suspicious schooners in Hampton Roads; and Captain McCauley returned to his home in Washington, receiving notice three weeks later from the Department confirming his detachment as commandant at Norfolk and consigning him to the oblivion of "wait orders."[2]

* * *

And still, nothing happened.

Not until 25 July 1861, three months after Paulding abandoned Gosport, did the federal government focus its attention on the loss of the navy yard. And it was the United States Senate that did so. On that date, a resolution was passed appointing a select committee with a specific mandate:

> To inquire into the circumstances attending the surrender of the navy yard at Pensacola, and the destruction of the property of the United States at the navy yard at Norfolk, and at the armory at Harper's Ferry ... and to inquire especially if there was any default in either case on the part of any officer of the United States.[3]

The force behind the Senate resolution was Senator John P. Hale of New Hampshire, chairman of its Committee on Naval Affairs, a politician who possessed a singular knack for criticism and a burning dislike of his fellow New Englander, Gideon Welles. Secretary Welles later put his finger on Hale's motive: "He desired to embarrass and assail the Navy Department. Nothing would have afforded him higher gratification than to have found the Secretary ... remiss and delinquent."[4] The other committee member—there were just two—was Iowa's Senator James Grimes, one of the best-informed navalists in Congress.

But the select committee tarried until November to hear its first testimony, interviewed only sixteen witnesses over a period of three weeks, then dragged out issuing its final report until 18 April 1862, over a month *after* the Battle of Hampton Roads and almost a year to the day since the Gosport Navy Yard had been burned. Despite its partisan origins and convoluted gestation, the report delivered a creditable evaluation of the events that led to the loss of the Gosport Navy Yard. But before delving into its findings, some reflections on the apparent disinterest of the Navy Department to initiate its own inquiries must be aired.

* * *

Writing years after the Civil War, naval historian Frank Bennett compiled the fruits of his labor to "get at the real truth" about Gosport, an affair he described as "utterly lacking in redeeming features." During the course of his research, he communicated with many of its participants. The core of his findings was prefaced by an uncomfortable truth:

> The principal officers concerned in the event were called upon to testify before the Senate Committee which investigated the Conduct of the War, and while they told the truth so far as they went, they told no more than was necessary, for at that time it would not have been either patriotic or politic, to have made some of the details public; and this restriction applies to a considerable extent even yet.[5]

Bennett's statement gives the strong hint that the Navy Department clamped down on any unnecessary proceedings that might expose the extent of its disaster at Gosport, allowing only that which was "politic" to reach the court of public opinion. Supporting this line of thought are Commodore Paulding's two known public statements validating the rationale for his actions that night of 20 April 1861 and which were published in national newspapers.[6] Paulding's accounts mirrored the Navy's official version of the events at Norfolk.

It might also explain another mystery. Within two weeks after the fall of the Pensacola Navy Yard, its commandant James Armstrong requested "that a court of inquiry may be instituted to enquire into all the facts and circumstances of said capture."[7] But a search of the *Official Records* fails to find any such similar requests by McCauley or Paulding.[8] Paulding's reticence is understandable; he immediately returned to his high-profile duties in the Office of Detail, in touch daily with Secretary Welles and never flinched from defending his snap decision to burn Gosport. But McCauley's situation was very different. He never again was recalled to active duty and, in time, became reviled as the epitome of naval disfunction and labeled by Welles as "feeble and incompetent."[9] The atmosphere of innuendo took its toll on McCauley, causing him to feel "his honor as an officer had been wounded, his reputation blemished."[10]

Why, then, didn't McCauley request a court of inquiry to vindicate his honor? Here we enter the realm of speculation. Could it be that Secretary Welles' ambiguous "applause" given at the end of his Armstrong court-martial general order satisfied McCauley's urge for a formal hearing? Or did he make the request only to have it unofficially denied, his letter returned unopened by a publicity-shy Navy Department? Or did he fear that any demand for official redemption could backfire, as it did with Armstrong, and rob him not only of his last shreds of dignity but also of his captain's pay?

It must have been hard on McCauley, waiting as the months ticked by till November and the opening of the Senate committee's hearings. In August he would have heard of Captain Paulding's appointment to the Navy's prestigious Ironclad Board, which a month later gave its blessing to John Ericsson's shot-proof battery, *Monitor*. And by October, Paulding was once again a commodore, this time in charge of the New York Navy Yard. Of Flag Officer Pendergrast, McCauley undoubtedly read the stories of *Cumberland*'s capture of prize after prize on the waters of Hampton Roads, and then of his appointment in September as the commandant of the Philadelphia Navy Yard. And he must have learned that Francis Gregory, next on the captain's list above him and long idled on "wait orders," had been recalled to superintend the Navy's "ninety-day gunboat" emergency shipbuilding program.[11]

McCauley couldn't escape running into his former colleagues and hearing the latest gossip—his Washington F Street home was but two blocks from the Navy Department, a few minutes' walk from the office that had ruled his life for the past fifty-two years. Head of a prominent family, he would have endured questioning stares at social gatherings. And during those empty months, as his fellow naval officers fought to preserve the Union, Captain Charles McCauley sat idle hoping for the vindication that might still be his.

* * *

The Senate's select committee assembled at the Capitol in early November 1861 and on the 7th recorded its first testimony. That same day, far down the eastern seaboard at Port Royal Sound, Flag Officer Samuel Du Pont's fleet won a resounding victory over Confederate forces, giving a boost to both the Navy and the nation.[12] But the joyous news had not yet filtered back to Washington when Paymaster John Bradford, late of the Gosport Navy Yard's receiving ship *Pennsylvania*, took his seat in the room of the Senate Committee on Naval Affairs and was sworn in.

Fifteen witnesses followed Bradford, the testimony portion of the committee's work spanning but three weeks, wrapping up on 26 November. The parade of personages was largely predictable with a few notable exceptions. Complimenting Bradford was Edward Doran, paymaster of the navy yard, and Marine Lieutenant Colonel Edelin. These three were the only officers originally attached to Gosport interviewed by the committee, silently confirming the magnitude of the Southern officers' resignations. The three complicit commodores, McCauley, Paulding and Pendergrast, all provided testimony, much of it riveting but also tedious in turns by the insertion of official orders. Of those officers who accompanied Paulding's expedition, only Lieutenants Wise and Russell and Marine Major

Nicholson testified, Wise's account providing the most arresting imagery. Engineer-in-Chief Isherwood was the sole officer to testify of those dispatched by Welles to remove *Merrimack* from the yard. Five navy yard employees and one resident of Portsmouth rounded out the committee's list of witnesses.

Two secretaries of the navy were of special interest to the committee, Secretary Isaac Toucey of the Buchanan administration and Secretary Gideon Welles of the current Lincoln administration. Only Secretary Toucey sat before the committee. Secretary Welles later explained his refusal to appear, his rationale commingled with his loathing for committee chairman Senator Hale:

> So far as the Department could furnish them, answers should be given to all specific enquiries, and that every facility should be extended to the Committee; but for myself I declined any appearance or explanation. My time, I assured the honorable chairman, was much too occupied in attending to necessary public duties ... I neither sought or shunned him. The records of the Department were thrown open to him, and they were a defense and justification.[13]

In total sixteen witnesses testified before the Senate committee. But the one man who held the knowledge of the decisions behind the Navy Department's every response to the threatening crises that overwhelmed President Lincoln's government kept silent. Absent their chief's executive-level explanations, his officers' testimony could expand on the narratives surrounding key events but revealed nothing of the motivations that precipitated them.

Extracts of the various witnesses' testimony have been extensively quoted throughout these pages. One last snippet is offered in closing, given by the witness with the most at stake, Captain McCauley. The final question posed at the end of his session was if "any efforts were made by the ... government ... prior to the 16th of April 1861, to reinforce you?" to which McCauley replied, "None were made during that time to my knowledge." He then added an unsolicited statement—and bear in mind that by the date of his testimony, it was well known that the Confederates were "making the *Merrimack* iron-clad:"[14]

> I desire to observe here, that as great stress seems to be laid upon the value of the *Merrimack*, I admit that, if fully equipped, with an efficient engine, she would have been a fine specimen of a ship-of war; but she was a dismantled hulk, with only her lower masts and

> bowsprit in, and without a particle of rigging to support her masts. The engine, which was in a disabled state, was hastily patched up; a few days of coal and provisions were put on board; there was no time to put her battery on board, and in no contingency could she have proceeded to sea in her then condition. As a matter of dollars and cents, the *Germantown* was worth two of her.[15]

* * *

Senate Report No. 37 was finally completed and sent to the printers on 18 April 1862. It totaled 123 pages of which thirty-seven were appendices. Of the two ancillary subjects identified in the committee's original scope, namely the surrender of the Pensacola Navy Yard and the destruction of the armory at Harpers Ferry, it accepted that they had already been satisfactorily dealt with by other official proceedings. Of its principal subject, the destruction and abandonment of the Gosport Navy Yard, its core findings were encapsulated within five main points.[16]

> 1st. The administration of Mr. Buchanan were guilty of neglect in not taking extraordinary care and employing every possible means to protect and defend this yard after indications of danger had manifested themselves.

Given what actually happened at Gosport, the Buchanan administration's most effective preventative measure would have been to begin the process of detaching and reassigning the yard's Southern officers. But few possessed the vision to recognize the necessity of doing so during those fraught antebellum months, and if they had, it is unlikely anyone could have mustered the courage or the political will to implement what would have been a sectionally explosive remedy.

> 2nd. The administration of Mr. Lincoln cannot be held blameless for suffering thirty-seven days to elapse after he came into power, before making a movement for the defense of the yard.

The observation offered above for the Buchanan administration applies here too, but with even greater urgency. To his credit, Secretary Welles attempted to dispatch batches of seamen to man the ships he ordered removed before 10 April (the thirty-seventh day), but was thwarted by external circumstances and his inability to insist on a higher priority for Gosport's needs. His handling of *Merrimack*'s removal, in hindsight, seems to have been motivated by the belief of his new engineer-in-chief, Isherwood, that if he could repair the steam frigate's machinery she would be saved. But power to the ship's screw could not overcome the fact that the man sent to sail her away, Commander Alden, was impotent without his own team of deck officers and sailors not to mention the cooperation of Gosport's officers—then *Merrimack* might have had a chance. Welles' irredeemable blunder came in the wake of learning that McCauley's officers had resigned and the yard was at great risk: he failed to inform McCauley that Paulding's rescue mission was on its way. It would have made all the difference. Taken as a whole, Welles' performance during his opening weeks as secretary of the navy was marred by his halting political leadership and imperfect exercise of naval command. And beyond the granular detail of defending Gosport, the Lincoln administration can be faulted for implementing a strategy to confront secession that overlooked the need to secure this vital navy yard.[17]

> 3rd. Captain McCauley was highly censurable for neglecting to send the *Merrimack* from the yard as he was ordered, and also for scuttling the ships and preparing to abandon the yard before any attack was made or seriously threatened, when he should have defended it and the property intrusted to him, repelling force by force, as he was instructed to do if the occasion should present itself.

After digesting the committee's opinion, it can be grasped why McCauley sensed the need to add his final statement regarding *Merrimack*'s worth. In pinning the blame on McCauley for declining to send away *Merrimack*, the committee apparently chose to ignore the risks to her reaching safety due to her being undermanned and inadequately prepared or from attempting to navigate an obstruction-strewn channel controlled by insurgents. The committee's second conclusion is also questionable. After the fact, it was easy to say that McCauley's decision to scuttle the ships was premature. But his determination to defend Gosport to the best of his ability was repeatedly vouched for by witnesses.[18] History also forgets that he handed over to Commodore Paulding an intact dockyard still under the control of the U.S. Navy. McCauley had faults to be sure. He may have been weak and was certainly unimaginative. And yes, he was fooled by his Southern officers, but no more so than Lincoln was by Virginia's secession or Welles was by the key players in the Forts Sumter-Pickens fiasco. Perhaps McCauley's biggest lapse centered on his failure to effectively communicate the precariousness of his situation—he should have been far more vocal in requesting aid from Secretary Welles.

> 4th. Captain Paulding was censurable for neglecting to consult with Captain McCauley and Captain Pendergrast while he was at the

> yard on the 17th of April, in regard to the course to be pursued in the event of an attack on the yard, and also for immediately, upon his arrival at the yard on the 20th, ordering the property to be burned and the yard abandoned before making proper means to satisfy himself that any necessity for such measures existed.

The committee correctly identified Paulding's two principal failures. His first inspection of the Gosport Navy Yard was perfunctory and evinced no foresight as to its impending crisis. Paulding's second chance to turn things around, his *Pawnee* expedition, again exhibited the same faulty, rushed thinking only on a grander scale. As for the yard's fate, he should have spread the decision-making among the senior captains present—McCauley, Pendergrast, Wilkes and Marston—and at the very least, waited until dawn to visually confirm that the dire remedy of destruction and abandonment was his only option. Further, Paulding should have been faulted for failing to seize the initiative and employ his two steamers, *Pawnee* and the tug *Yankee*, to take control of the Elizabeth. Decisive and bold action might have kept Norfolk in Union hands.

> 5th. Captain Pendergrast, in command of the *Cumberland*, the flag-ship of the home squadron, made no suggestions as to the measures proper to be adopted, and seems to have taken no part in the transaction, except to move his ship as directed.

Again, the committee displayed its perception by identifying Pendergrast's lack of initiative. As flag officer of the Home Squadron, he failed to aggressively use all resources afloat at his disposal to prevent the obstruction of the Elizabeth's main channel. His former record while in command of *Merrimack* during her first commission demonstrated flawed leadership—he did not run a happy ship. Pendergrast's stilled voice in the face of Gosport's worsening crisis lends credence to whispers hinting he shrank from active participation in deference to his wife and her Barron family secessionist sentiments. His share of accountability for Gosport's loss has never been adequately recognized: McCauley consulted with Pendergrast before making all his major decisions.

* * *

Six days after the Senate committee heard its final witness, Gideon Welles presented his first annual report of the secretary of the navy to President Lincoln. Welles' report embraced the whole of the Navy's operations, from the accomplishments of its squadrons to matters of regulation, enlistment and budget. Throughout its many pages of detail, he never directly addressed the surrender of the Pensacola and Norfolk Navy Yards or the consequences to the Navy of their loss. He mentioned them only once and then in a brief passing reference:

> The yards at Norfolk and Pensacola, being in the hands of the insurgents, the chief of the Bureau of Yards and Docks had not had them under his supervision but for a small portion of the year. His labors, however, have not been diminished or less arduous in consequence.[19]

When the Senate committee's report on the loss of the Norfolk Navy Yard finally became available in late April 1862, Navy officers must have eagerly devoured it and none more so than Flag Officer Samuel Du Pont. Still on board his flagship *Wabash*, anchored in the sheltered waters of Port Royal, he offered his opinions on what he had just read in a journal entry to his wife Sophie:

> Mr. Bridge sent me the Senate report of the Naval Committee on the loss of the Norfolk Navy Yard and the destruction of property there. It is very severe and damaging on Commodore McCauley, in reference to whom there can be no difference of opinion; if a person of his feebleness could be held as a responsible being, in England or in France he would have lost his life—he would have been hung like Byng. But the committee are also very severe on Commodore Paulding and Pendergrast, the later not so much. The truth is the loss of that navy yard was such a blow to us in consequences and such a gain to the rebels that I make some allowance for the severity of the Committee. But Paulding in the *Evening Post*, in a letter of great ability and tact, makes a manly defense of himself.[20]

* * *

A year after the traumatic events that overwhelmed the Gosport Navy Yard, a verdict on its principal participants' behavior had been delivered by the Congressional committee. In the months, years and decades that followed, the lives of those involved ineluctably moved on, stamping yet more judgments on how they were remembered. What did fate have in store for them?

Commodore Pendergrast was the first to pass from the scene. While serving as commandant of the Philadelphia Navy Yard, he was felled by a paralytic stroke in November 1862, one month before his sixtieth birthday. His wife Susan, Captain Barron's daughter, lived on for another eleven years.[21]

Captain McCauley was formally placed on the Navy's retired list on 21 December 1861, a month after he testified before the Senate committee. Five years later he was promoted to the rank of commodore, although during the intervening years, he was never recalled to active duty. He died in Washington on 21 May 1869 at the age of seventy-six. An extract from his obituary, making reference to the "affair" surrounding the Norfolk Navy Yard, read as follows:

> The Congressional Committee appointed to investigate the affair failing to exonerate him entirely from blame in the matter, he deemed that his honor as an officer had been wounded, his reputation blemished—the effect of which was to plunge him into the deepest melancholy, causing disease of the heart, of which he died ... The funeral of Commodore McCauley, on Sunday, May 22, 1869, was one of the most imposing ceremonies witnessed in Washington for years.[22]

This same obituary noted that his venerable uncle, Admiral Charles Stewart of *Constitution* fame, "yet lives" at the age of ninety-two.

Of the three commodores, Hiram Paulding was the last to go. A rear admiral at the war's end, he served as the governor of the Naval Asylum in Philadelphia and finally as Boston's port admiral. The end finally came at his home in Huntingdon, Long Island. He died in October 1878 at age eighty of embolism of the heart. His obituary was headed "A Hero Gone to His Rest."[23]

Secretary Welles learned quickly on the job. His native abilities helped him become one of the best secretaries of the navy of the nineteenth century. He unapologetically defended the Navy's handling of the Gosport Navy Yard affair until the end. Welles served two full terms before losing his job to the incoming Grant administration. He died at his home in Hartford in February 1878 at age seventy-six.[24]

Commander Alden possessed a sterner character than that displayed at Gosport. He commanded ships during the fierce engagements of the passage of New Orleans' forts and the Battle of Mobile Bay. Rising to the rank of rear admiral, Alden lived until 1877, dying at age sixty-six.[25]

Engineer Isherwood was appointed the first chief of the new Bureau of Steam Engineering in 1862, and held that position until Admiral Porter dismissed him in 1869, an unfortunate consequence of the ingrained line officers versus staff animosity. Isherwood finally retired in 1884 but continued to actively publish his research until near the end of his life. He died in 1915, the year after World War I began, at age ninety-two. Many consider him the greatest engineer ever produced by the U.S. Navy.[26]

The commodores may have been wise to deny Lieutenant Selfridge temporary command of *Dolphin* to guard the Elizabeth's channel. During his very active Civil War career, he survived the sinking of three ships, the last two of which he commanded, all of their names beginning with "C": the sloop *Cumberland*, the ironclad *Cairo* and the timberclad *Conestoga*. Finally retiring as a rear admiral, he hauled his flag down only months before the beginning of the Spanish-American War. Of all the players in the drama surrounding Gosport Navy Yard, Selfridge was the last to leave the stage. He died two days short of his eighty-eighth birthday in 1924.[27]

Many of the yard's Southern officers who resigned also experienced adventuresome careers. Ordnance Lieutenant "Terry" Sinclair turned up in Scotland where he organized the building of a clandestine cruiser on the model of the famous *Alabama*, whose sinking off Cherbourg he later witnessed. His superior at Gosport, Commander John Tucker, went on to command Charleston's squadron of ironclads before the fall of that city forced him ashore, after which his naval brigade fought on until surrendering at Saylor's Creek. His postwar experience included seven years under the Peruvian flag exploring the Amazon. The eloquent Commander Richard Page's Confederate service culminated in his late 1864 heroic defense of Mobile Bay's Fort Morgan. After the war, he returned to Norfolk where among other pursuits he superintended the city's public schools. He died in 1901 at age ninety-three.[28]

And what became of Virginia's vocal insurrectionary, Governor Henry Wise? Soon after the war began Wise was appointed a brigadier general in the Confederate Army. During the ironclad *Virginia*'s construction, several of her officers were guests at his nearby home, where Wise hinted it was he who first proposed her design. Later in the war, General Wise helped defend Petersburg and was with Lee when he surrendered at Appomattox. Unrepentant to the end, he refused to sign the ironclad oath. Henry Wise lived in Richmond during his last years, passing away in 1876 at age sixty-nine.[29]

As for the Gosport Navy Yard, federal forces recaptured it on 10 May 1862, a little over a year after it had been destroyed and abandoned. As of that date, the yard was officially redesignated the U.S. Navy Yard, Norfolk, Virginia, wiping out the duality, confusion and perhaps shame that clung to its old name.[30] Ten days later Commander John W. Livingston, who had reported for duty to Commodore McCauley the day before Paulding ordered the yard burned, was appointed the renamed navy yard's first commandant.[31]

A new era had begun in the long and storied life of the Norfolk Navy Yard.

* * *

APPENDIX 1

Annotated Calendar, March and April 1861

An overview of the key months of March and April 1861 is presented in this appendix, a calendar's graphic format serving to emphasize the rush of events leading up to the destruction and abandonment of the Gosport Navy Yard. Only forty-eight days elapsed between Lincoln's inauguration and the immolation of the United States Navy's premier dockyard.[1]

The four months preceding March 1861 were perhaps even more tumultuous to the national cohesion of the United States of America. Following Lincoln's election on 6 November 1860, South Carolina swiftly moved to secede, passing its ordnance of secession on 20 December. Early in the new year, it was followed by six other deep South states. By 18 February 1861, Jefferson Davis had been inaugurated as the president of the Confederate States of America. Seven of America's former states had now formalized their disunion and congealed to form a new republic. And two more weeks had yet to pass before Abraham Lincoln took his oath of office.[2]

* * *

TABLE A1.1 *Annotated Calendar, March and April 1861*

MARCH.						
Sun.	Mon.	Tues.	Wed.	Thurs.	Fri.	Sat.
...	...	...	...	...	1	2
3	4 Lincoln Inaugurated	5 Welles sworn in as SECNAV	6	7	8	9
10	11 *Germantown* ready by 25th	12 Armstrong CM convened	13	14	15	16
17	18	19	20	21	22	23 *Cumberland* at Hampton Roads
24	25	26	27	28	29 *Cumberland* orders to Norfolk	30 Welles orders ships for Sumter
31 300 men ordered to Norfolk	...	...	...	...	...	...
APRIL.						
Sun.	Mon.	Tues.	Wed.	Thurs.	Fri.	Sat.
...	1	2	3	4 VA Convention votes for Union	5	6
7	8	9	10 Q: how long to ready *Merrimack*	11 Order to prepare *Merrimack*	12 Fort Sumter bombarded	13 Fort Sumter surrenders
14 Isherwood starts work at Gosport	15 Lincoln calls up 75,000 militia	16 Paulding ordered to visit Gosport	17 VA votes for secession	18 *Merrimack* ready McCauley holds	19 VA militia show up; *Pawnee* sails	20 Ships scuttled; Paulding arrives
21 Gosport burned and abandoned	22	23	24 Armstrong CM verdict Gen Ord	25	26	27
28	29	30	...	...	...	...

* * *

APPENDIX 2

List of Officers at the Gosport Navy Yard

The pages of tables following present a detailed listing of U.S. Navy officer presence at the Gosport Navy Yard during the critical period of Virginia's secession and the yard's destruction and abandonment. A close examination of the first table reveals the organizational structure of the navy yard in the opening months of 1861. A core staff denoted navy-yard managed the basic repair and shipbuilding work fundamental to a dockyard. Smaller subgroups, ordnance and hospital, looked after matters of gunnery and medicine, respectively. The receiving ship *Pennsylvania* held the yard's reservoir of unassigned crew, mostly seamen but also firemen and coal heavers. The officer detailed to ordinary-minded the ships that were out of commission and in ordinary. Rendezvous staff actively recruited new sailors or engine room hands, and the marine guard provided the yard's basic security. The civil officers, officially appointed to their various positions but not commissioned or warranted as naval officers, rounded out the yard's key personnel.

The more complex tables that follow were created to graphically display each officer's loyalty during the six days of crisis that culminated in the yard's destruction and abandonment. The heavy, dark horizontal bars indicating resignation or dismissal clearly demonstrate how the Navy's management team was effectively decapitated by the sudden departure of its Southern officers. Note the effect of Virginia passing her ordinance of secession on 17 April. With the exception of Lieutenant Alexander Semmes, the navy yard's originally appointed line officers had disappeared by 20 April. More ominously all its ordnance officers "went South." Only the receiving ship *Pennsylvania* retained some organizational coherence despite the fact that her captain and executive officer had resigned just days before the yard's abandonment.

The state listed after each officer's name was the state of his citizenship. In most cases, it was also the state of his birth and appointment, but that clean assignment didn't always hold true. Commander John Tucker considered an ardent Virginian, was born in and remained a citizen of the District of Columbia, but was appointed from Indiana, married to a Virginian woman and, in his later years, lived in Norfolk. And although Marine Captain Jabez Rich hailed from Maine, he threw his lot in with the rebels. However, for nearly all officers, the given state of citizenship was a reliable guide to their ultimate allegiance, Northerners to the Union and Southerners to their home state.

Most Southern officers who resigned close to the date of the yard's destruction were not accorded the privilege of being listed as such but were officially documented by the Navy as "dismissed," an irrevocable strike against a man who had pledged his sword to his nation.

Significantly, the officers assigned to the yard in the days immediately prior to Gosport's abandonment—Livingston, Parrott and Irwin—were all from Northern states, a clear indication that Secretary Welles had grasped the metastasizing danger of Virginia's secession. The one exception was Naval Constructor John Porter, who had in early April applied to replace the reassigned Samuel M. Pook. Porter shrewdly waited until the yard's new owners, the Commonwealth of Virginia, had taken over before he resigned his appointment as a U.S. naval constructor. He went on to be the progenitor of the South's ironclad steam ram *Virginia.*

* * *

TABLE A2.1 *Officers at the Gosport Navy Yard*[1]

NAVY-YARD

Captain Charles S. McCauley, Commandant.

Commander, Robert G. Robb.

Lieutenants, Charles F. M. Spotswood, Alex. A. Semmes, and William Sharp.

Surgeon, R. T. Mason.

Paymaster, Edward C. Doran.

Paymaster, John Johnson, Inspector of Provisions.

Boatswain, William Smith.

Gunner, John Owins.

Carpenter, James Meads.

Sailmaker, William N. Maull.

ORDNANCE

Commander, John R. Tucker.

Lieutenant, George T. Sinclair.

Gunner, Chas. B. Oliver, Magazine Keeper.

HOSPITAL

Surgeon, Samuel Barrington.

Passed Assistant Surgeon, Dinwiddie B. Phillips.

RECEIVING SHIP, PENNSYLVANIA

Commander, Arthur Sinclair.

Lieutenants, Carter B. Poindexter, Edward Donaldson and John Watters.

Surgeon, John Rudenstein.

Passed Assistant Surgeon, Charles H. Williamson.

Paymaster, John O. Bradford.

Carpenter, John T. Rustic.

Sailmaker, George Thomas.

ORDINARY

Commander, Thomas R. Rootes.

RENDEZVOUS

Commander, Richard L. Page.

Lieutenant, Peter U. Murphey.

MARINE GUARD

Lieutenant Colonel, James Edelin.

Captain, Jabez Rich (on receiving ship).

First Lieutenant, Adam N. Baker.

CIVIL OFFICERS

Navy Agent, George Loyall.

Naval Store-keeper, Robert M. Boykin.

Naval Constructor, Samuel M. Pook.

Civil Engineer, Calvin Brown.

* * *

TABLE A2.2 *Gosport Navy Yard Officer Presence During Virginia's Secession and the Yard's Destruction*

Officer Name and Rank	State	16 April	17 April	18 April	19 April	20 April	21 April
Yard	*Citizen of*						
CAPT Charles S. McCauley, Commandant	PA						
CDR Robert G. Robb	VA	Earlier det		Resigned			
CDR John W. Livingston (relieved Robb)	NY				Reported		
LT Charles F. M. Spotswood	VA	Earlier det		Resigned			
LT Enoch G. Parrott (relieved Spotswood)	NH					Reported	
LT Alexander A. Semmes	MD						
LT William Sharp	VA		Dismissed				
Surgeon Randolph T. Mason	VA						Later dism
Paymaster Edward C. Doran	IN			To Wash			
Paymaster John Johnson	NC					Dismissed	
Chief Engineer Robert Danby	DE				Spec duty		
Bos'n William Smith	VA					Dismissed	
Gunner John Owins	VA					Dismissed	
Carpenter James Meads	VA					Dismissed	
Sailmaker William N. Maull	PA						
Ordnance							
CDR John R. Tucker	DC			Resigned			
LT George T. Sinclair	VA	Detached		Dismissed			
Gunner Chas. B. Oliver	VA						Dismissed
Hospital							
Surgeon Samuel Barrington	PA	Earlier det					
Pass Asst Surgeon Dinwiddie B. Phillips	VA						Later resn
Receiving Ship Pennsylvania							
CDR Arthur Sinclair	VA			Resigned			
LT Carter B. Poindexter	VA			Resigned			
LT Edward Donaldson	MD						
LT John Watters	MI	Detached					
LT John Irwin (relieved Watters)	PA					Reported	
Surgeon John Rudenstein	MD						
Pass Asst Surgeon Charles. H. Williamson	VA						Later dism
Paymaster John O. Bradford	PA	Earlier det					

TABLE A2.3 *Gosport Navy Yard Officer Presence, cont'd.*

Officer Name and Rank	State	16 April	17 April	18 April	19 April	20 April	21 April
Pennsylvania, cont'd							
Carpenter John T. Rustic	VA						Later dism
Sailmaker George Thomas	n.a.						
Ordinary							
CDR Thomas R. Rootes	VA				Resigned		
Rendezvous							
CDR Richard L. Page	VA			Resigned			
LT Peter U. Murphey	NC						Resigned
Marine Guard							
LtCol James Edelin	MD						
Capt Jabez Rich	ME					Joined reb	
1st Lt Adam N. Baker	PA						Later dism
Civil Officers							
Navy Agent George Loyall		Earlier res					
Naval Storekeeper Robert M. Boykin							
Naval Constructor John L. Porter	VA		Reported				Later resn
Civil Engineer Calvin Brown							

NOTES:

1. Earlier det = Detached at an earlier date than the table range from assigned duty.
2. Earlier res = Resigned at an earlier date than the table range from assigned duty.
3. Detached = Detached on that date from assigned duty.
4. Resigned = Resigned commission from U.S. Navy on that date. In most cases, dismissal from service recorded on same date.
5. Reported = Reported for assigned duty on that date.
6. Dismissed = Dismissed from U.S. Navy on that date.
7. Later dism = Dismissed from U.S. Navy on a later date than the table range.
8. Later resn = Resigned from U.S. Navy on a later date than the table range.
9. Joined reb = Joined Rebel forces on that date.
10. Remained in service to the U.S. Navy:
11. Resigned or dismissed from service to the U.S. Navy:

APPENDIX 3

Secretary Welles' Orders of 16 April 1861

Presented in this appendix are two orders issued by Secretary Welles on 16 April 1861, one addressed to Commodore McCauley, commandant of the Gosport Navy Yard, and the second to Commodore Pendergrast, flag officer of the Home Squadron. Both of these letters were hand-carried to Norfolk by Commodore Paulding during the course of his first flying visit to the navy yard.

A full discussion of each order is given in Chapter 9. Note in particular the nuanced balance of Welles' language, alternating between specific demands by the Department and conditional suggestions, the combination of which gave wide room for interpretation:

> [Confidential.] NAVY DEPARTMENT, April 16, 1861
>
> SIR: The events which have transpired since my confidential communication to you of the 10th instant impose additional vigilance and care in protecting the public property under your charge, and placing the vessels and stores, if necessary, beyond jeopardy.
>
> Referring to my letter of the 10th, you will continue to carry out the instructions therein contained. The Engineer-in-Chief, B. F. Isherwood, who was dispatched to Norfolk to aide in putting the *Merrimack* in condition to be moved, reports that she will be able to take her departure on Thursday. It may not be necessary, however, that she should leave at that time, unless there is immediate danger pending. But no time should be lost in getting her armament on board, and you will also place the more valuable public property, ordnance, stores, etc., on shipboard, so that they can at any moment be moved beyond the reach of seizure. With diligence on your part, it is not anticipated that any sudden demonstration can be made which will endanger either the vessel or stores. The

Plymouth and *Dolphin* should, however, be placed beyond danger of immediate assault at once, if possible. The *Germantown* can receive on board stores and ordnance from the yard, and be towed out by the *Merrimack* if an assault is threatened. Men have been ordered from New York to man and assist in moving the vessels, but recent demands have left an insufficient number to meet the requisition. Under these circumstances, should it become necessary, Commodore Pendergrast will assist you with the men from the *Cumberland.* You will please to submit this letter and my confidential communication of the 10th to Commodore Pendergrast, who will assist and cooperate with you in carrying the views of the Department into effect. As it is difficult at this distance to give instructions in detail, the Department has thought proper to dispatch Commodore Paulding to Norfolk, who will be the bearer of this communication, and explain to yourself and Commodore Pendergrast the views and purposes of the Department. You will be pleased to advise with him freely and fully as to your duties and the interests of the Government in the present threatening emergency. The vessels and stores under your charge you will defend at any hazard, repelling force by force, if necessary, any and all attempts to seize them, whether by mob violence, organized effort or any assumed authority.

I have the honor to be, very respectfully, yours, etc.,

GIDEON WELLES

Secretary Navy.

Commodore C. S. McCauley,

Norfolk, Va.[1]

* * *

[Confidential.] NAVY DEPARTMENT, April 16, 1861

Sir: A state of things has arisen which renders the immediate departure of the *Cumberland* as originally intended inexpedient. Events of recent occurrence, and the threatening attitude of affairs in some parts of our country, call for the exercise of great vigilance and energy at Norfolk. Confidential communications have been heretofore made to Commodore McCauley on these subjects, which he will submit to you; and Commodore Paulding, who brings this letter to you, will verbally and more in detail explain the views of the Department. Please to advise freely and fully with both these gentlemen, and cooperate with them in defending the vessels and public property at the navy yard. As there is an insufficiency of men in the service at that station for moving vessels, it may become necessary to render assistance from the force under your command.

Until further orders, the departure of the *Cumberland* to Vera Cruz will be deferred. In the meantime you will lend your assistance, and that of your command, toward putting the vessels now in the yard in condition to be moved, placing ordnance and ordnance stores on board for moving, and in case of invasion, insurrection, or violence of any kind to suppress it, repelling assault by force if necessary. The *Cumberland* can render effective service, and it is deemed fortunate that the Government is enabled to avail itself of your service and that of your command at this juncture at Norfolk.

I am, sir, respectfully, your obedient servant,

GIDEON WELLES,

Secretary Navy.

Commodore G. J. Pendergrast,

Commanding U.S. Sloop Cumberland, Norfolk, Va.[2]

* * *

APPENDIX 4

Fate of Warships at the Gosport Navy Yard

This appendix presents a snapshot of the fate of the Navy's warships at Gosport following its destruction and abandonment. Of all the eleven ships listed, only one of them was in commission as a U.S. Navy warship, the receiving ship *Pennsylvania. Germantown, Plymouth* and *Dolphin* were nearly ready for sea; although armed they were still out of commission and unmanned. *Merrimack* was being actively repaired. The remaining five vessels were afloat on the Elizabeth River in "ordinary" except for the ship-of-the-line *New York*, which had yet to be completed and was in frame on the building ways under Ship House A.

When a Navy warship was taken out of commission for overhaul, repair or long-term layup, she was considered to be in "ordinary." In addition to lowering her flag and discharging her officers and crew, a warship in ordinary was substantially stripped of all of her ordnance, stores and equipment. Her guns and shot were hoisted ashore and her powder was sent to the yard's magazine. All stores, victuals and water barrels were removed from her hold; sails, rigging and tackle were taken to the sailmaker and rigger's lofts, and her upper masts, spars and yards were all sent down and stored on shore. If she was a steamer, her bunkers were emptied of coal. In many cases, the upper deck was roofed over with a rudimentary structure to protect the ship's interior from the elements. A caretaker crew under the supervision of the officer in charge of ordinary would make periodic checks of each ship's condition, pumping the bilges and inspecting her moorings. It can be appreciated that preparing a ship in ordinary for active service at sea required a considerable amount of time and effort, taking anywhere from ten days to over a month, depending on the ship's condition. Even a vessel "ready for sea" but still under the yard's care and out of commission needed a massive effort by her officers and crew before she could truly be capable of sustaining any significant operations.

The list of men-of-war also makes clear the wide variety of warship classes that were in the care of the Gosport Navy Yard. Mighty ships-of-the-line, big frigates, smaller sloops-of-war and a brig-of-war were all afloat at the yard. Even a steamer, the double-banked steam frigate *Merrimack*, was included in their numbers. In *Merrimack*'s case, her machinery had been condemned and was, as resources and funding allowed, being slowly replaced. The Navy's fleet at Gosport ranged from the very valuable *Merrimack*, a powerful steamer only six years old, to the useless relic *United States*, the Navy's first super frigate built at the close of the previous century. As varied as the eleven listed warships were, both in class and condition, by count they constituted twelve percent of the entire United States Navy of 1861. But by their number of guns, they represented a far more powerful twenty-six percent of the Navy's total firepower afloat. And all of them were doomed.

* * *

TABLE A4.1 *Fate of Warships at the Gosport Navy Yard, 21 April 1861*

Name When Built[1]	Guns and Class[2]	Location in Yard[3]	Fate[4]
Merrimack 1855	40-guns Steam Frigate	Alongside wharf, south of ship houses	Scuttled, burned to waterline; later *Virginia*
Pennsylvania 1837	120-guns Ship-of-the-Line	In the stream, north at yard's entrance	Burned to the water's edge
Columbus 1819	80-guns Ship-of-the-Line	In the stream, upper yard	Scuttled, otherwise uninjured
Delaware 1820	84-guns Ship-of-the-Line	In the stream, upper yard	Scuttled, otherwise uninjured
Columbia 1836	50-guns Frigate	In the stream, off St Helena pier	Burned to the water's edge
Raritan 1843	50-guns Frigate	In the stream, south of *Columbia*	Burned to the water's edge
United States 1797	50-guns Frigate	In the stream, south of *Raritan*	Abandoned uninjured; later used as receiving ship
Germantown 1846	22-guns Sloop-of-War	Alongside wharf, under sheers	Injured by falling sheers, much damaged by fire
Plymouth 1843	22-guns Sloop-of-War	Alongside wharf	Scuttled and partially burned, later raised
Dolphin 1836	4-guns Brig-of-War	Alongside wharf	Scuttled, burned to the floor heads
New York stocks	84-guns Ship-of-the-Line	Inside Ship House A	Burned, totally destroyed

NOTES:

1. Of all the ships listed above, only the yard's receiving ship, *Pennsylvania*, was in commission. *Merrimack* was alongside the wharf under repair; *Germantown, Plymouth* and *Dolphin* were nearly ready for sea but as yet unmanned; and *New York* was incomplete and still on the stocks. All the remaining vessels were in ordinary.
2. Two other commissioned U.S. Navy warships were present at Gosport when it was abandoned: *Cumberland*, a 24-gun sloop-of-war serving as Flag Officer Pendergrast's Home Squadron flagship, and *Pawnee*, a 4-gun steam sloop-of-war attached to the Home Squadron which had just arrived carrying Commodore Paulding and the officers and men attached to his expedition. Together with the commercial steam tug *Yankee*, they all escaped destruction.

* * *

APPENDIX 5

Destination of the Guns Seized from the Gosport Navy Yard

In the words of Admiral David Dixon Porter, "Great as was ... the loss of our ships, it was much less than the loss of our guns."[1] Gosport was the repository of huge numbers of naval guns, most removed from the many warships laid up within its waters. The complete battery of a two deck ship-of-the-line could number as many as eighty-four pieces, a frigate's weapons added another fifty guns to the count. It was from this treasure of heavy naval guns that the South gained its greatest bounty from the abandonment of the Gosport Navy Yard. Porter later wrote that the Navy encountered Gosport's purloined guns on nearly every navigable waterway they ventured upon, be it along the coast, at harbor entrances, on inland sounds and bays or far up the innumerable Southern rivers.

Strangely enough, the yard's commandant, Charles Stewart McCauley, and the man sent to rescue the yard and who then decided to destroy it, Hiram Paulding, had only an inflated notion of the actual number of naval guns in their charge. McCauley testified to the Senate select committee enquiring into the loss of Gosport, "I think nearly three thousand heavy pieces of ordnance." were in the yard when it was captured. Commodore Paulding opined "I suppose there were between two and three thousand." Data later submitted by Andrew Harwood, chief of the Bureau of Ordnance and Hydrography, presumably accessing substantially complete records, provided what should have been the definitive figure. "The inventory ... shows that there were in the yard at the time of its destruction 768 guns of all calibres."[2] However, as will be seen, this statement veered in the opposite direction and grossly underestimated the number of captured cannons.

Paymaster William H. Peters, newly commissioned by the Virginia Navy on the day after the yard's abandonment, soon presented his new masters with a report tallying all the public property left behind by the fleeing federal forces. He counted a total of 1,085 guns, which probably included those taken from armed U.S. Navy vessels burned during the yard's destruction, namely *Pennsylvania, Germantown, Plymouth,* and *Dolphin* (*Merrimack* had no guns on board), thus accounting for some of the increase over Harwood's number.[3]

Testimony at the Senate hearings into Gosport's loss by yard machinist James H. Clements offered a more nuanced explanation: "the ordnance bureau charging the guns to the destination of the ship instead of to the place where the guns were being fitted."[4] Thus, in the case where a ship's battery was reduced, the extra guns landed would remain in the yard but still be counted as part of the ship's original ordnance outfit.

The figure for the number of cannons seized at Gosport that comes closest to an agreement with the many and varied sources is Admiral Porter's estimate of "at least twelve hundred fine guns."[5]

Reading Peters' report shows why Porter so lamented the loss of Gosport's guns: they spanned the range of all standard U.S. Navy ordnance of the day. Several hundred modern Dahlgren 9-inch and 8-inch guns headed the list which was fleshed out by the ubiquitous 42, 32 and 24-pounders of the era and other smaller calibers down to a bronze 9-pounder. Substantial numbers of shell guns, 8-inch and even a 10-inch, were also included as were 9-inch naval guns of a type unspecified.[6] These were all powerful weapons that would have taken the agrarian South years to import or manufacture. It was indeed a lethal harvest that Paymaster Peters' colleagues distributed throughout the Confederacy.

And where were all of Gosport's guns sent? According to Peters' "Inventory Report," over sixty percent of the weapons were distributed to numerous batteries in the Norfolk area and beyond, to locations along the James River, in Richmond and throughout the cities and states of the Confederacy. Those listed were shipped from 21 April through 1 July 1861, or during the period when Virginia controlled the navy yard prior to handing it over to the Confederate government. In nearly all cases, the guns were shipped "with carriages, etc., complete."

An early beneficiary of the batch of guns sent to New Orleans was Captain Raphael Semmes. One 8-inch gun and four 32-pounders arrived in time to arm his first commerce raider, *Sumter*, which sailed from the Head of the Passes on 30 June 1861.[7] Surely the strangest entry appears at the end—forty guns to a Captain Thomas in Yankee-occupied Baltimore. The tables below give a breakout of the destinations of the bounty that Admiral Porter came to lament.

TABLE A5.1 *Destinations of the Guns Seized from the Gosport Navy Yard*[8]

NORFOLK AND THE ELIZABETH RIVER

Battery at Norfolk City, 3 guns
" Naval Hospital, 15 guns
" Naval Magazine, Fort Norfolk, 8 guns
" Pinner's Point, 7 guns
" Lambert's Point, 6 guns
" Craney Island, 30 guns
" Bousch's Bluff, 5 guns
" Sewell's Point, 14 guns

AREA SURROUNDING NORFOLK

Battery at Pig Point, 14 guns
" Powell's Point, 4 guns
Kempsville, 2 guns
Seaboard & Roanoke Railroad, 1 gun

RICHMOND, JAMES RIVER AND VIRGINIA

Richmond, 136 guns
Richmond, for General Beauregard, 16 guns
City Point, 1 gun
Battery at Burwell's Bay, 6 guns
Fredericksburg, for Captain R. G. Robb, 4 guns

CONFEDERATE CITIES AND STATES

Charleston, South Carolina, to Gen. Beauregard, 39 guns
Savannah, Georgia, 10 guns
Pensacola, Florida, 10 guns
New Orleans, Louisiana, to Commo. Rousseau, 23 guns
Memphis, Tennessee, to R. P. Pugh, 5 guns
Tennessee, 32 guns

SPECIFIC COMMANDERS OR THEATERS

General Gwynn, to North Carolina, 52 guns

" " " " , 45 guns

" " " " , 90 guns

" " to Fort Macon, 18 guns

Lieut. George T. Sinclair, and by him sent to Army South, 34 guns

Captain Thomas at Baltimore, Maryland, 40 guns

* * *

APPENDIX 6

The Potential of Conversions to a Steam Frigate

It is all too easy for historians to dismiss the value of the ships-of-the-line and sailing frigates that were scuttled and burned when the Gosport Navy Yard was abandoned. Lying for many years in "Rotten Row," floating high on the water, their paint peeling and stripped of rigging, these old ships would seem to be obsolete relics rather than useful war vessels. However, their stout oak hulls had considerable life left in them if reconfigured to embrace the latest developments in naval technology. As the following examples show, plans for doing so had been actively considered by the Navy during the years before Gosport's destruction and abandonment.

John Lenthall, chief of the Bureau of Construction, Equipment and Repairs, had realized the old ships' potential and outlined a proposed conversion of *Columbus*, an 80-gun liner then lying in ordinary at Gosport. See the attached letter to Secretary of the Navy Isaac Toucey dated 18 August 1857.[1] Lenthall's scheme involved cutting away the upper deck from the hull ("razeeing" the ship) and installing steam machinery; the resulting war vessel would have been an auxiliary steam frigate of the largest class. As he pointed out, similar schemes had been carried out by European navies. It was even possible, Lenthall suggested, to pay for the conversion with funds already available from existing appropriations.

Congress also got in on the act, adding a proviso to its naval appropriations bill of 22 June 1860, directing the secretary of the navy to appoint a board "to examine the condition of the sailing vessels of the Navy, and the cost of giving them ... full steam power together with the expediency of making such a change, in view of the cost, condition, model and general character of such vessels so altered." Lenthall and Chief Engineer Benjamin Isherwood were among those appointed to the board.[2] In his final annual report of December 1860, Secretary Toucey recommended that eight of the Navy's old liners be converted into steam frigates, and that five of its sailing frigates be razeed into efficient sloops of war (similar to the conversion applied years before to *Cumberland*).[3]

While such conversions might not have been immediately useful in prosecuting the brown water naval conflict that followed, they would have proven their worth if America's Civil War had assumed a European dimension through the intervention of England or France. Their ready availability for alteration into sea-going cruisers—or possibly even ironclads—might have made all the difference to a desperate Union facing not only a Southern rebellion but also an oceanic war.

* * *

Navy Department
Bureau of Const. &c
August 18th 1857

Sir:

I would respectfully submit for consideration the subject of the repair and alteration of the Ship-of-the-line "Columbus."

There are several ships-of-the-line afloat, all of which require very extensive repairs and the manner in which it can be best executed and most to the interest of the service, seem to be a question of some moment.

The quantity of useful material in these ships is large and of much value whilst in shape, but if taken to pieces will be worth but little.

They can be rebuilt as ships-of-the-line or converted into auxiliary steamers as has been done with many such vessels in the English and French navies.

The "Columbus" being of smaller size than the other ships cannot be made formidable as a ship-of-the-line, but might become so if razeed or converted into a frigate of the very largest class with a light upper deck or as an auxiliary steam frigate, but without the upper deck.

The conversion of a ship of this class is more advantageously executed than with one of less rate on account of there being relatively greater available space and the removal of an entire deck, the reducing the spars and also by reducing the number of guns by giving a greater space between them which is much needed, and consequently diminishing the number of men, quantity of provisions and water necessary for them, will, if not altogether, nearly equal the weight of the machinery to be added, if too great a sacrifice is not made to speed under steam alone.

If the Department considers the subject favorably, a more full examination of the ship by a board to consist of a naval officer, the naval constructor and chief engineer of the Gosport yard where the ship is, will more fully show if there be the advantages above named.

If the work be not pressed forward too rapidly, the appropriation on hand will permit the trial of the system, and the steam machinery can be made in that yard, bringing the expenditure altogether under control, requiring no further outlay than the state of the appropriation will warrant.

I am, Sir, very respectfully
Your Obt. Servant
John Lenthall
Chief of the Bureau

* * *

NOTES

CHAPTER 1 *Gosport Navy Yard*

1 37th Congress, 2nd Session, 18 April 1862, Senate Report No. 37, 46, McCauley's testimony. Hereinafter cited as Senate Report No. 37. Also see NARA, RG 24, M330, Roll 9, Captains—Active List, 8 Charles S. McCauley, "July 25/60 To the Command of Navy Yard at Norfolk;" Ibid., Captains—Reserved List, 150 Charles Stewart, "Resumed command of the Phila Navy Yard on the 20th Sept 1859." Hereinafter cited as NARA, M330, Roll number, date and subject.

2 Mileage data was taken from the lower left-hand margin of Map No. 6, *Panorama of the Seat of War, Bird's Eye View of Virginia, Maryland, Delaware and the District of Columbia*, 1861. Sea route distances between the cited points would be somewhat longer.

3 John V. Quarstein and Dennis Mroczkowski, *Fort Monroe, The Key to the South* (Charleston, etc.: Arcadia Publishing, 2000), 8-9.

4 Marshall W. Butt, *A Brief History of Norfolk Naval Shipyard, Portsmouth, Virginia* (Portsmouth: Public Information Office, 1956), 1-2. Hereinafter cited as Butt, *Norfolk.*

5 Edward P. Lull, *History of the United States Navy-Yard at Gosport, Virginia (near Norfolk)* (Washington: Government Printing Office, 1874), 8-9. Hereinafter cited as Lull, *Gosport.*

6 Ibid., 11-18. Donald L. Canney, *Sailing Warships of the US Navy* (Annapolis: Naval Institute Press, 2001), 46-48, 199. Hereinafter cited as Canney, *Sailing Warships.*

7 Lull, *Gosport*, 16-17, Article II of an act passed by the Virginia legislature authorizing sale of the Gosport Navy Yard.

8 Spencer C. Tucker and Frank T. Reuter, *Injured Honor, the Chesapeake-Leopard Affair, June 22, 1807* (Annapolis: Naval Institute Press), 1996.

9 *Appleton's Cyclopedia of American Biography*, James Grant Wilson and John Fiske, Eds. (New York: D. Appleton and Company, 1888), Vol. IV, 78. Hereinafter cited as *Appleton's*, McCauley.

10 Claude Berube and John Rodgaard, *A Call to the Sea, Captain Charles Stewart of the USS Constitution* (Washington, D.C.: Potomac Books, Inc, 2006), 68-70. Hereinafter cited as Berube, *A Call to the Sea.*

11 Lull, *Gosport*, 23.

12 *Appleton's*, McCauley. Appleton's identifies the British frigate as *Narcissus.*

[13] Benson J. Lossing, *Lossing's Pictorial Field Book of the War of 1812* (New York: Harper & Brothers, 1868), Vol. II, 675-680. Another must for understanding the true contours of this American victory is Christopher Pieczynski, "The Decision at Craney Island," American Battlefield Trust at www.battlefields.org website, accessed February 2022.

[14] Canney, *Sailing Warships*, 95-104, 200.

[15] Lull, *Gosport*, 26-36.

[16] Ibid., 39-42.

[17] Ibid., 44-45. Also see Donald L. Canney, *Lincoln's Navy, The Ships, Men and Organization, 1861-65* (Annapolis: Naval Institute Press, 1998), 48-50. Hereinafter cited as Canney, *Lincoln's Navy.*

[18] NARA, M330, Roll 9, Captains—Active List, 8 Charles S. McCauley.

[19] Lull, *Gosport*, 45-46. Also see Senate Report No. 37, various testimony.

[20] *Navy Register*, 1861, 100.

[21] Ibid., adjusted for known detachments and assignments up to April 1861.

[22] Claude Berube, *On Wide Seas, The US Navy in the Jacksonian Era* (Tuscaloosa: The University of Alabama Press, 2021), 26. The cited quote was from Admiral Steedman's memoirs.

[23] Drawn from various sources including David P. Werlich, *Admiral of the Amazon: John Randolph Tucker, His Confederate Colleagues and Peru* (Charlottesville and London: University of Virginia Press, 1990), hereinafter cited as Werlich, *Admiral of the Amazon*; Chris Michael, *Lelia* (Birkenhead: Countryvise Limited, 2004), hereinafter cited as Michael, *Lelia*; Alan B. Flanders, *John L. Porter, Naval Constructor of Destiny* (Whitestone, VA: Brandylane Publishers, Inc. 2000), hereinafter cited as Flanders, *John L. Porter.*

CHAPTER 2 *Charles Stewart McCauley*

[1] Berube, *A Call to the Sea*, 3.

[2] *Lamb's Biographical Dictionary of the United States*, John Howard Brown, Ed. (Boston: Federal Book Company of Boston, 1903), Vol. V, 195. Hereinafter cited as *Lamb's*, McCauley.

[3] NARA, M330, Roll 3, McCauley, Chas. S, "Mids. Rank 16 Jan. 1809."

[4] NARA, RG 36, M1880, Roll 8, Charles Stewart McCauley, 1807.

[5] "The Late Commodore McCauley, United States Navy," *The New York Times*, 23 May 1869. Hereinafter cited as McCauley Obituary, *NY Times*. Also see "Commodore Charles Stewart McCauley, U.S.N." Boston *The Pilot*, 31 July 1869. Hereinafter cited as McCauley Obituary, Boston *Pilot.*

[6] James E. Bradford, Ed., *Command Under Sail, Makers of the American Naval Tradition 1775-1850* (Annapolis: Naval Institute Press, 1985), 205-06; Lull, *Gosport*, 19-22.

[7] *Lamb's*, McCauley.

[8] NARA, M330, Roll 3, MCawley Chas S. [sic].

[9] McCauley Obituary, *NY Times*; McCauley Obituary Boston *Pilot.*

[10] *Appleton's,* McCauley; *Lamb's,* McCauley. *Jefferson*'s hull was discovered in the 1980s beneath a marina at Sackets Harbor, revealing much about Chauncey's ships.

[11] NARA, M330, Roll 3, MCawley Chas S. [sic].

[12] Ibid., M'Cauley, Daniel S.

[13] Ibid., MCawley Chas S. [sic].

[14] *Lamb's,* McCauley.

[15] NARA, M330, Roll 3, M'Cauley Charles S.

[16] Robert Irwin Johnson, *Thence Around Cape Horn* (Annapolis: United States Naval Institute, 1963), 27-30, 221.

[17] NARA, M330, Roll 3, M'Cauley Charles S. The two twelve-month furlough orders were dated 22 December 1822 and 1 November 1823.

[18] *Appleton's,* McCauley.

[19] NARA, M330, Roll 3, M'Cauley, Daniel S, "Resignation accepted 13 April 1825."

[20] Ibid., Roll 3, M'Cauley Charles S. He was briefly "ordered from N York to Decoy" but this was altered two weeks later before he was finally sent to Boston in early August.

[21] Ibid., Roll 4, Charles S. McCauley, "Decr 28 1825. Rept to Capt Crane for the Boston."

[22] *Lamb's,* McCauley.

[23] NARA, M330, Roll 4, Charles S. McCauley, "July 22 1829 Leave Unlimited."

[24] Paul H. Silverstone, *The Sailing Navy, 1775-1854* (Annapolis: Naval Institute Press, 2002), 61. Hereinafter cited as Silverstone, *Sailing Navy.*

[25] NARA, M330, Roll 4, Charles S. McCauley, "3 March 1831, commission dated, promotion on 12 March 1831."

[26] Ruth Ketring Nuermberger, "Asbury Dickins (1780-1861): A Career in Public Service." *The North Carolina Historical Review,* Vol. XXIV, No. 3 (July 1947): 281-314. Hereinafter cited as Nuermberger, "Dickins."

[27] Ibid.

[28] NARA, M330, Roll 5, Chas. S. McCauley, "Feby 12, 1833 det from the Rendez Balto & appointed to Comd the Fairfield."

[29] Ibid., "May 4th, detd from Comd of Fairfield at his request."

[30] Ibid., "Oct 22d 1834, appointed to the comd of the St Louis."

[31] *Lamb's,* McCauley; NARA, M330, Roll 5, Chas. S. McCauley, "Aug 4th leave 3 mos."

[32] Nuermberger, "Dickins," 300-313.

[33] NARA, M330, Roll 5, Chas. S. McCauley, "March 6th/37 to N Yard Phila."

[34] Berube, *A Call to the Sea,* 216.

[35] Find A Grave, "Como Charles Stewart McCauley," https://www.findagrave.com/memorial/37275570/charles-stewart-mccauley, accessed July 2021. Hereinafter cited as Find A Grave, "McCauley."

[36] NARA, M330, Roll 6, Chas. S. McCauley, "9th December 1839."

[37] Stephen Chapin Kinnaman, *Captain Bulloch, The Life of James Dunwoody Bulloch, Naval Agent of the Confederacy* (Indianapolis: Dog Ear Publishing, 2013), 91-104. Bulloch served as a midshipman on *Delaware* during McCauley's three-year cruise.

[38] NARA, M330, Roll 6, Chas. S. McCauley, "March 13th/44 detd & lv 3 mos."

[39] Find A Grave, "McCauley."

[40] NARA, M330, Roll 7, Charles S. McCauley, "July 27th/46 to comd Washington yard on 1st Sept."

[41] Taylor Peck, *Round-Shots to Rockets, A History of the Washington Navy Yard and U.S. Naval Gun Factory* (Annapolis: United States Naval Institute, 1949), 105-107.

[42] Find A Grave, "McCauley."

[43] NARA, M330, Roll 7, Charles S. McCauley, "April 18/50 To Command the Pacific Squadron."

[44] Ibid. Also see the *Navy Register*, 1850-53.

[45] "The President in Earnest about Cuba—An Apology will be Insisted on—War Probable," Brooksville *Indiana American*, 20 April 1855, reprint of article from *The New York Times*; NARA, M330, Roll 7, Charles S. McCauley, "April 6/55 Apptd to Command Home Squadron."

[46] *Appleton's*, "McCauley."

[47] NARA, M330, Roll 7, Charles S. McCauley, "June 14/55 Detd & apptd Member of Efficiency Board."

[48] Charles Oscar Paullin, *Paullin's History of Naval Administration* (Annapolis: United States Naval Institute, 1968), 238-42. Hereinafter cited as Paullin, *History*.

[49] NARA, M330, Roll 7, Charles S. McCauley, "May 7/56 Duty ... to prepare system of Watch & Station Bills;" Ibid., Roll 9, Charles S. McCauley, "Special Duty Washington;" *Lamb's*, McCauley.

[50] Eighth Census of the United States (1860), Washington, D.C.

[51] NARA, M330, Roll 9, Francis H. Gregory, Charles S. McCauley, "Sea Service" data.

CHAPTER 3 *The Coming Ordeal*

[1] David M. Potter, *The Impending Crisis, 1848-1861* (New York: Harper & Row, Publishers, Inc., 1976), 343, 405-447. Hereinafter cited as Potter, *Impending Crisis*.

[2] Ibid., 490-99.

[3] Paolo E. Coletta, Ed., *American Secretaries of the Navy, Volume 1 1775-1913* (Annapolis: Naval Institute Press, 1980), 315. Hereinafter cited as Coletta, *Secretaries*.

[4] Potter, *Impending Crisis*, 515-22.

[5] Geoffrey C. Ward with Ken Burns and Ric Burns, *The Civil War* (New York: Alfred A. Knopf, Inc., 1990), 48.

[6] Paullin, *History*, 250; *Navy Register*, 1861.

[7] Potter, *Impending Crisis*, 540-41.

[8] *ORN*, I, 4, 5. Secretary Toucey to Commodore Armstrong, 3 January 1861.

[9] Ibid., 6, Secretary Toucey to Captain Glynn, 5 January 1861; Ibid., 9, Secretary Toucey to Flag Officer Pendergrast, 9 January 1861.

[10] Canney, *Lincoln's Navy*, 50.

[11] Maury Klein, *Days of Defiance, Sumter, Secession, and the Coming of the Civil War* (New York: Alfred A. Knopf, 1997), 205-07. Hereinafter cited as Klein, *Days of Defiance.*

[12] NARA, M330, Roll 9, James Armstrong, "13 Sept/60 to Command Navy Yard Pensacola."

[13] *Navy Register*, 1861.

[14] *ORN* I, 4, 21-38. Testimony from Captain Armstrong's court of enquiry, 8 to 15 February 1861.

[15] Ibid., 33. The naval constructor was John L. Porter, who would latter gain fame as the designer of *Virginia* and many other Confederate ironclads.

[16] Ibid., 31, Testimony by William Musgrave, assistant engineer, 14 February 1861. Notable among the officers with secessionist tendencies were the yard's executive officer, Commander Farrand, Lieutenant Renshaw and storekeeper Samuel Gonzales.

[17] Ibid., 7, Commodore Armstrong to Secretary Toucey, 8 January 1861.

[18] Potter, *Impending Crisis*, 542-43.

[19] *ORN*, I, 4, 220. Secretary Toucey to Captain Barron, 7 January 1861; Secretary Toucey to Captain Walker, 7 January 1861.

[20] *Samuel Francis Du Pont, A Selection from His Civil War Letters*, John D. Hayes, Ed. (Ithaca, NY: Cornell University Press, 1969), I, 19, letter to Mrs. Du Pont, 13 January 1861. Hereinafter cited as Du Pont, *Civil War Letters.*

[21] *ORN*, I, 4, 221. Captain Walker to Secretary Toucey, 15 January 1861.

[22] Ibid., 34-36, Testimony by Navy Lieutenant John Irwin, 14 February 1861, and Slemmer's Second Lieutenant Gilman, 15 February, 1861; Ibid., 47, Captain Storer's concluding statement at Armstrong's court of enquiry, 20 February 1861. Also see Klein, *Defiance*, 205-07, for a colorful account of the company's hurried transfer to Fort Pickens.

[23] Ibid., 16-17, Captain James Armstrong to Secretary Toucey, 25 January 1861.

[24] Klein, *Days of Defiance*, 207.

[25] *ORN*, I, 4, 16-17, Captain Armstrong to Secretary Toucey, 25 January 1861.

[26] Ibid., 18, Captain Armstrong to Secretary Toucey, 26 January 1861.

[27] Ibid., 272, Secretary Toucey to Captain Powell, 24 January 1861.

[28] Senate Report No. 37, 58, Toucey's testimony.

[29] Ibid., 46, McCauley's testimony.

[30] *ORN*, I, 4, 274, Secretary Welles to Commodore Breese, 31 March 1861. In this case, while sailors were ordered to Norfolk, the name of the ship(s) on which they were to serve was not given.

[31] James E. Valle, *Rocks and Shoals, Naval Discipline in the Age of Fighting Sail* (Annapolis: Naval Institute Press, 1980), 50-54.

[32] *ORN*, I, 4, 19, Secretary Toucey to Captain Storer, 2 February 1861.

[33] Ibid., 47-48, Court findings as read by Captain Storer, 20 February 1861.

[34] Potter, *Impending Crisis*, 547.

CHAPTER 4 *Old Dominion's Unwanted Decision*

[1] Potter, *Impending Crisis*, 545-46.

[2] Ibid.

[3] *Showdown in Virginia, The 1861 Convention and the Fate of the Union*, William W. Freehling and Craig M. Simpson, Eds. (Charlottesville and London: University of Virginia Press, 2010), x. Hereinafter cited as Freehling, *Showdown*.

[4] Potter, *Impending Crisis*, 545-47.

[5] Freehling, *Showdown*, xi-xiii.

[6] Ibid., 49-50, George Wythe Randolph's secessionist speech, March 16.

[7] Encyclopedia Virginia, "John Letcher (1813-1884)," https://encyclopediavirginia.org/entries/letcher-john-1813-1884/, accessed December 2022.

[8] Craig M. Simpson, *A Good Southerner, The Life of Henry A. Wise of Virginia* (Chapel Hill and London: The University of North Carolina Press, 1985), 239-40. Hereinafter cited as Simpson, *A Good Southerner*.

[9] Ibid., 3. Also see Barton H. Wise, *The Life of Henry A. Wise of Virginia* (New York: The Macmillan Company, 1899), 15. Hereinafter cited as Wise, *Henry Wise*.

[10] Simpson, *A Good Southerner*, 18, 115, 175.

[11] Wise, *Henry Wise*, 63-65.

[12] Simpson, *A Good Southerner*, 159. Also see Encyclopedia Virginia, "Henry A. Wise (1806-1876)," https://encyclopediavirginia.org/entries/wise-henry-a-1806-1876/, accessed December 2022. Hereinafter cited as Encyclopedia Virginia, "Wise."

[13] Wise, *Henry Wise*, 172.

[14] Ibid., 108, 119.

[15] Simpson, *A Good Southerner*, 210; Encyclopedia Virginia, "Wise."

[16] Wise, *Henry Wise*, 262-63; Simpson, *A Good Southerner*, 222.

[17] Wise, *Henry Wise*, 265-66.

[18] Simpson, *A Good Southerner*, 235-36.

[19] Wise, *Henry Wise*, 267-68.

[20] Simpson, *A Good Southerner*, 238.

[21] Ibid., 241.

[22] Wise, *Henry Wise*, 268. A quote from Wise in a letter dated 1 December 1860.

[23] Potter, *Impending Crisis*, 546-50.

[24] Ibid., 549-51.

[25] Ibid., 551.

CHAPTER 5 *Lincoln Takes Command*

[1] Klein, *Days of Defiance*, 312.

[2] Potter, *Impending Crisis*, 566-68.

[3] Wise, *Henry Wise*, 386. Letter by Fernando Wood to Henry Wise dated 19 February 1866.

[4] Paullin, *History*, 251. Also see Coletta, *Secretaries*, I, 320-361. The section on Welles was written by John Niven, the author of a comprehensive biography of Welles, *Gideon Welles, Lincoln's Secretary of the Navy* (Baton Rouge and London: Louisiana State University Press, 1973). Hereinafter cited as Niven, *Welles*.

[5] Paullin, *History*, 250.

[6] Ibid., 253-54.

[7] Ibid., 210-11; *Navy Register*, 1860-1861. Also see Virginia Places, "Before the Pentagon, Where Were the Military Headquarters?" http://www.virginiaplaces.org/military/beforethepentagon.html, accessed February 2019.

[8] *Navy Register*, 1861.

[9] Stephen Chapin Kinnaman, *Merrimack, The Biography of a Steam Frigate* (Wilmington and Malaga: Vernon Press, 2019), 15-20. Hereinafter cited as Kinnaman, *Merrimack*.

[10] *Navy Register*, 1861. Also see John H. Schroeder, *Shaping a Maritime Empire, The Commercial and Diplomatic Role of the American Navy, 1829-1861* (Westport and London: Greenwood Press, 1985), 139-64.

[11] Canney, *Lincoln's Navy*, 9-10.

[12] Milestones in the History of U.S. Foreign Relations, "French Intervention in Mexico and the American Civil War, 1862-1867," https://history.state.gov/milestones/1861-1865/French-intervention, accessed December 2022.

[13] Gideon Welles, *Diary of Gideon Welles, Secretary of the Navy under Lincoln and Johnson*, 3 Vols. (Boston and New York: Houghton, Mifflin Company, 1911), I, 10. Hereinafter cited as Welles, *Diary*.

[14] Niven, *Welles*, 325.

[15] William S. Dudley, *Going South: U.S. Navy Officer Resignations & Dismissals On The Eve Of The Civil War* (Washington, D.C.: Naval Historical Foundation, 1981), 6, 16-17, Table II. Hereinafter cited as Dudley, *Going South*.

[16] Coletta, *Secretaries*, I, 324.

[17] Dudley, *Going South*, 13, 21.

[18] Gideon Welles, *Civil War and Reconstruction*, Albert Mordell, Ed. (New York: TWayne Publishers, 1959), 31. A reprint of the article: Gideon Welles, "Mr. Welles in Answer to Mr. Weed, The facts of the Abandonment of the Gosport Navy-Yard," *The Galaxy*, X (July 1870): 15-35. Hereinafter cited as Welles, *Civil War*.

[19] *ORN*, I, 4, 53. From the text of a naval general order issued on 24 April 1861. One researcher found that "the proceedings of Captain Armstrong's court-martial appears to be missing from the rest of the proceedings deposited in the National Archives." See West Gulf Squadron, "General Naval Order," National Republican, Washington DC, 4 May 1861, www.westgulfsquadron.net/articles/generalorder.html, accessed December 2022.

[20] Welles, *Civil War*, 22.

[21] Ibid., 23.

[22] Ibid.; Welles, *Diary*, I, 41.

[23] Welles, *Civil War*, 23.

[24] *ORN* I, 4, 272. Commodore McCauley to Secretary Welles, 11 March 1861.

[25] Welles did eventually request 250 men be sent to Norfolk, but did so three weeks later, long after they were needed there on the 25th. See *ORN* I, 4, 273, Secretary Welles to Commodore Breese, 31 March 1861.

[26] Potter, *Impending Crisis*, 562-63; Niven, *Welles*, 326.

[27] J. Thomas Scharf, *History of the Confederate Navy* (New York: Rogers and Sherwood, 1887), 128. Hereinafter cited as Scharf, *Confederate Navy*. Also see Paullin, *History*, 254.

[28] Welles, *Diary*, I, 34.

[29] Ibid., 13.

[30] Ibid., 14-15. Also see Ari Hoogenboom, *Gustavus Vasa Fox of the Union Navy* (Baltimore: The Johns Hopkins Press, 2008), 60. Hereinafter cited as Hoogenboom, *Fox*.

[31] Welles, *Diary*, I, 13-14.

[32] Potter, *Impending Crisis*, 575-76.

CHAPTER 6 *Cumberland Arrives at Hampton Roads*

[1] *ORN*, I, 4, 266, Flag Officer Pendergrast to Secretary Welles, 23 March 1861. The narrative of *Cumberland*'s arrival at Hampton Roads was borrowed, with alterations, from a similar description of her anchoring off Vera Cruz in 1846, as observed by her chaplain F. W. Taylor and quoted in Canney, *Sailing Warships*, 74-75.

[2] *ORN*, I, 4, 76, Secretary Toucey to Flag Officer Pendergrast, 2 February 1861.

[3] Canney, *Sailing Warships*, 74-76. Also see Senate Report No. 37, 32, 37 for her crew size and the *Navy Register*, 1861, 102, for her captain and officers.

[4] Kinnaman, *Merrimack*, 29-74.

[5] "Vessels at the Charlestown Navy-Yard," *The New York Times*, 8 August 1856.

[6] NARA, RG 45, M125, Roll 356, Captain Pendergrast to Secretary Dobbin, 20 August 1856.

[7] Kinnaman, *Merrimack*, 30-31.

[8] Klein, *Days of Destiny*, 379.

[9] Senate Report No. 37, 40.

[10] *ORN*, I, 4, 266, Flag Officer Pendergrast to Secretary Welles, 25 March 1861.

[11] Ibid., 273, Commodore McCauley to Secretary Welles, 26 March 1861.

[12] *Navy Register*, 1861, 16-17.

[13] *ORN*, I, 4, 267, Flag Officer Pendergrast to Commodore McCauley, 27 March 1861.

[14] Ibid., 268, Secretary Welles to Flag Officer Pendergrast, 29 March 1861; Ibid., Flag Officer Pendergrast to Secretary Welles, 31 March 1861.

[15] See Chapter 20 for an overview of the Senate committee's origins and hearings. A summary of its work, Senate Report No. 37, was issued on 18 April 1862, a year after Gosport's loss.

[16] Senate Report No. 37, 38.

[17] Career data of the Gosport Navy Yard officers was taken from the *Navy Register*, various years; NARA, M330, Roll 9; Lewis Randolph Hamersly, *The Records of Living Officers of the U.S. Navy and Marine Corps* (Philadelphia: J. B. Lippincott & Co., 1870), hereinafter cited as Hamersly, *Records*; and other sources as specifically noted. Also see the Naval History and Heritage Command (NHHC) website, "Officers—Continental & U.S. Navy/Marine Corps 1775-1900," https://www.history.navy.mil/research/library/online-reading-room/title-list-alphabetically/o/officers-continental-usnavy-mc-1775-1900.html, accessed July, 2022. Hereinafter cited as NHHC website, "Officers."

[18] Werlich, *Admiral of the Amazon*, 1-22.

[19] Michael, *Lelia*, 6.

[20] *Lamb's Biographical Dictionary of the United States*, John Howard Brown, Ed. (Boston: Federal Book Company of Boston, 1903), Vol. VI, 108. Hereinafter cited as *Lamb's*, Page.

[21] Senate Report No. 37, 92, Attachment No. 5, List of officers of the expedition … .

[22] Flanders, *John L. Porter*, 37-38.

[23] Welles, *Civil War*, 24.

[24] *ORN*, I, 4, 227, President Lincoln to Secretary Welles, 29 March 1861.

[25] Welles, *Diary*, I, 53; Welles, *Civil War*, 26.

[26] *ORN*, I, 4, 228, Secretary Welles to Commodores Buchanan, McCauley and Breese, 30 March 1861.

[27] Ibid., 273, Secretary Welles to Commodore Breese, 31 March 1861.

[28] Welles, *Diary*, I, 42.

[29] Ibid., I, 3.

[30] *ORN*, I, 4, 269, Superintendent Blake to Secretary Welles, 15 April 1861. Blake referred to his earlier suggestion made on the 5th. Also see Ibid., 274-75, Secretary Welles to Commodore McCauley, 11 April 1861.

[31] Ibid., 274, Secretary Welles to Commodore McCauley, 10 April 1861.

[32] Ibid., 227, Secretary Welles to Captain Mercer, 28 March 1861; Ibid., 229, Secretary Welles to Commodore Breese, 1 April 1861.

[33] Ibid., 265, Captain Mercer to the Secretary of the Navy, 13 March 1861; "The Note of Preparation," *The New York Times*, 6 April 1861.

[34] Klein, *Days of Defiance*, 383.

[35] ORN, I, 4, 234, Secretary Welles to Commodore Foote, 5 April 1861; Ibid., 235, Secretary Welles to Captain Mercer, 5 April 1861.

[36] Welles, *Civil War*, 27.

[37] *ORN*, I, 4, 230-31, Commodore Foote to Secretary Welles, 2 April 1861.

[38] Welles, *Diary*, I, 42-43.

[39] Coletta, *Secretaries*, I, 326. The biographer was John Niven.

CHAPTER 7 *Virginia in the Balance*

[1] Freehling, *Showdown*, xiii.

[2] Ibid.; Potter, *Impending Crisis*, 492.

[3] Klein, *Days of Defiance*, 382.

[4] Freehling, *Showdown*, 154.

[5] Wise, *Henry Wise*, 272.

[6] Freehling, *Showdown*, 155-56.

[7] Ibid., 49, 155, 172.

[8] "The Virginia State Convention," *The New York Times*, 11 April 1861.

[9] Frank M. Bennett, *The Steam Navy of the United States* (Pittsburgh: Warren & Co., Publishers, 1896), 236. Hereinafter cited as Bennett, *Steam Navy*.

[10] NHHC Website, "Officers." Also see Edward William Sloan, III, *Benjamin Franklin Isherwood, Naval Engineer* (Annapolis: United States Naval Institute, 1965). Hereinafter cited as Sloan, *Isherwood*.

[11] Bennett, *Steam Navy*, 232. The author has taken the liberty of revising Bennett's spelling of *Merrimac* to the correct *Merrimack*.

[12] Ibid.

[13] Sloan, *Isherwood*, 189, 194-95.

[14] U.S. Naval Academy Museum, The Ives Collection, FICm.009.0015 (238-DSC-0661-62), B. F. Isherwood to John Lenthall, 19 January 1861.

[15] *ORN*, I, 4, 274, Secretary Welles to Commodore McCauley, 10 April 1861.

[16] One of the vexing aspects of the many letters and orders reproduced in the *Official Records* is that while the dates they were written were well documented, the date on which they were received was much less certain. Were they carried by hand, courier or put in the mail pouches?

[17] Lull, *Gosport*, 40.

[18] Bennett, *Steam Navy*, 232.

[19] *ORN*, I, 4, 280-81, Engineer-in-Chief Isherwood to Secretary Welles, 18 April 1861.

[20] Kinnaman, *Merrimack*.

[21] Senate Report No. 37, 50, McCauley's testimony.

[22] Welles, *Diary*, I, 43.

[23] *ORN*, I, 4, 274-75, Secretary Welles to Commodore McCauley, 11 April 1861.

[24] Ibid., 269-70, Superintendent Blake to Secretary Welles, 15 April 1861.

[25] Ibid., 275, Secretary Welles to Commodore McCauley, 11 April 1861.

[26] Ibid., 275, Secretary Welles to Commodore Breese, 11 April 1861.

[27] Ibid., 300-02, Commander Alden to Secretary Welles, 30 November 1861.

[28] Ibid., 275-76, Secretary Welles to Commander Alden, 11 April 1861.

[29] Welles, *Diary*, I, 44.

[30] Hamersly, *Records*, 36; NARA, M330, Roll 9, James Alden; NHHC Website, "Officers."

[31] Bennett, *Steam Navy*, 232.

[32] *ORN*, I, 4, 276, Secretary Welles to Commodore McCauley, 12 April 1861.

[33] Ibid., Secretary Welles to Engineer-in-Chief Isherwood, 12 April 1861.

[34] Freehling, *Showdown*, 170. Randolph exaggerated the quantity of gunpowder by a factor of ten. The navy yard magazine held no more than 3,000 barrels of powder.

[35] Ibid., 169, 171.

[36] Welles, *Diary*, I, 23. The excerpt was lifted from Welles' reference to getting *Powhatan* away from New York on 6 April, an episode laden with unexpected and unintended consequences.

CHAPTER 8 *Fort Sumter Surrenders*

[1] Potter, *Impending Crisis*, 578, 582.

[2] Klein, *Days of Defiance*, 378.

[3] Welles, *Diary*, I, 16-21; Potter, *Impending Crisis*, 580.

[4] *ORN*, I, 4, 109, President Lincoln to Lieutenant Porter, 1 April 1861; Ibid., President Lincoln to Captain Mercer, 1 April 1861; Ibid., President Lincoln to Commandant Navy Yard, N.Y., 1 April 1861.

[5] Klein, *Days of Defiance*, 387-88; Welles, *Diary*, I, 23-26.

[6] Welles, *Diary*, I, 26, 31.

[7] *ORN*, I, 4, 245-51, Assistant Secretary Fox to Secretary Welles, 24 February 1865.

[8] Ibid.

[9] Freehling, *Showdown*, 160.

[10] "The News in Washington," *The New York Times*, 13 April 1861.

[11] Freehling, *Showdown*, 160.

[12] Potter, *Impending Crisis*, 581.

[13] "A Proclamation by the President of the United States," *The New York Times*, 15 April 1861.

[14] Scharf, *Confederate Navy*, 136.

[15] Stephen Chapin Kinnaman, *John Lenthall, The Life of a Naval Constructor* (Wilmington and Malaga: Vernon Press, 2022), 247. Hereinafter cited as Kinnaman, *Lenthall*.

[16] *Navy Register*, 1861; Senate Report No. 37, 58, testimony by Secretary Toucey.

[17] NARA, M330, Roll 9, Robert Robb, "April 15/61, Det'd & wait orders;" Ibid., Charles Spotswood, "April 15/61, Det'd & to command the *Perry*."

[18] Ibid., John Watters, "April 16/61, Det'd & to the *Minnesota*;" Ibid., George T. Sinclair, "April 16/61, Det'd & wait orders."

[19] *ORN*, I, 4, 81-82, Report of Lieutenant Irwin, 10 February 1861; NARA, M330, Roll 9, John Irwin, "April 15/61 To the Receiving Ship at Norfolk."

[20] Freehling, *Showdown*, 161.

[21] Simpson, *A Good Southerner*, 248.

[22] Freehling, *Showdown*, 165.

[23] Simpson, *A Good Southerner*, 248.

[24] "The Course of Virginia," *The New York Times*, 15 April 1861.

[25] Wise, *Henry Wise*, 273-74.

[26] Simpson, *A Good Southerner*, 248.

[27] Freehling, *Showdown*, 166.

CHAPTER 9 *Merrimack—The Prize*

[1] Alexander Crosby Brown, *Steam Packets on the Chesapeake, A History of the Old Bay Line Since 1840* (Centreville, MD: Tidewater Publishers, 1961), 25-49.

[2] Ibid.

[3] William Howard Russell, *My Diary, North and South* (Boston: T. O. H. P. Burnham, 1863), 81.

[4] Ibid., 83.

[5] Ibid., 85.

[6] *ORN*, I, 4, 300, Commander Alden to Secretary Welles, 30 November 1861; Marshall W. Butt, *Portsmouth Under Four Flags, 1752-1970* (Portsmouth: Messenger Printing Co., Inc., 1971), 36.

[7] *ORN*, I, 4, 300, Commander Alden to Secretary Welles, 30 November 1861.

[8] Ibid., 280-81, Engineer-in-Chief Isherwood to Secretary Welles, 18 April 1861.

[9] Ibid.; Bennett, *Steam Navy*, 235; Kinnaman, *Merrimack*, 115-17.

[10] John W. H. Porter, *A Record of Events in Norfolk County, Virginia, from April 19th, 1861, to May 10th, 1862 ...* (Portsmouth, VA: W. A. Fiske, Printer and Bookbinder, 1892), 10. Hereinafter cited as Porter, *A Record*. The author was the son of Naval Constructor John Luke Porter, the man who supervised the building of the ironclad ram *Virginia*.

[11] Bennett, *Steam Navy*, 239.

[12] Kinnaman, *Merrimack*, 17, 31, 281; Canney, *Sailing Warships*, 199-200.

[13] Senate Report No. 37, 50, McCauley's testimony.

[14] *ORN*, I, 4, 301, Commander Alden to Secretary Welles, 30 November 1861. Although Commander Robb had been detached on 15 April, he was still on duty until his relief arrived.

[15] Senate Report No. 37, 50, McCauley's testimony.

[16] *ORN*, I, 4, 277, Commodore McCauley to Secretary Welles, 16 April 1861.

[17] Welles, *Civil War*, 24.

[18] Welles, *Diary*, I, 46.

[19] Hamersly, *Records*, 21-22.

[20] David F. Long, *Gold Braid and Foreign Relations, Diplomatic Activities of U.S. Naval Officers 1798-1883* (Annapolis: Naval Institute Press, 1988), 133-36.

[21] NARA, M330, Reel 9, Hiram Paulding, "April 4/61 ... to special duty in Detailing officers."

[22] *Autobiography of Rear Admiral Charles Wilkes, U.S. Navy 1798-1877*, William James Morgan, et al., Eds. (Washington: Naval History Division, Department of the Navy, 1978), 757. Hereinafter cited as Wilkes, *Autobiography*.

[23] *ORN*, I, 4, 277-78, Secretary Welles to Commodore McCauley, 16 April 1861.

[24] Ibid., 278, Secretary Welles to Commodore Pendergrast, 16 April 1861.

[25] Ibid., 278, Secretary Welles to Commodore Paulding, 16 April 1861.

[26] *Navy Register*, 1861, 16.

[27] Bennett, *Steam Navy*, 236-37.

CHAPTER 10 *Virginia's Agony, Wise's Insurrection*

[1] Freehling, *Showdown*, 167.

[2] Ibid.

[3] Simpson, *A Good Southerner*, 248.

[4] Ibid., 249.

[5] Freehling, *A Good Southerner*, 174, 179.

[6] John Sherman Long, "The Gosport Affair, 1861," *The Journal of Southern History*, Vol. 23, No. 2 (May 1957): 159. Hereinafter cited as Long, "Gosport Affair."

[7] Wise, *Henry Wise*, 274-75; Spencer C. Tucker, *Brigadier General John D. Imboden, Confederate Commander in the Shenandoah* (Lexington: The University Press of Kentucky, 2003), 15-20. Hereinafter cited as Tucker, *Imboden*.

[8] Wise, *Henry Wise*, 275. The brass field pieces were actually made of bronze.

[9] *Battles and Leaders of the Civil War*, Robert Underwood Johnson and Clarence Clough Buel, Eds. (New York: The Century Co., 1884-87), I, 112. The chapter was titled "Jackson at Harpers Ferry in 1861," by John D. Imboden Brigadier-General, C.S.A. Hereinafter cited as Imboden, *Battles and Leaders*.

[10] Wise, *Henry Wise*, 276-77.

[11] Ibid., 277.

[12] Senate Report No. 37, 78-79, Isherwood's testimony.

[13] Lull, *Gosport*, 53; *ORN*, I, 4, 279, Captain Marston to Flag Officer Pendergrast, 17 April 1861.

[14] Scharf, *Confederate Navy*, 132.

[15] Porter, *A Record*, 12.

[16] Senate Report No. 37, 46, McCauley's testimony; Ibid., 39-40, Pendergrast's testimony.

CHAPTER 11 *Paulding's Flying Visit*

[1] *ORN*, I, 4, 279, Captain Marston to Flag Officer Pendergrast, 17 April 1861; Senate Report No. 37, 40, Pendergrast's testimony. "I was on shore daily up to the time when the obstructions were sunk."

[2] William N. Still, Jr., Ed., *What Finer Tradition, The Memoirs of Thomas O. Selfridge, Jr., Rear-Admiral, U.S.N.* (Columbia: University of South Carolina Press, 1987), 27. Hereinafter cited as Still, *Selfridge*. Also see Senate Report No. 37, 46, McCauley's

testimony. "I took upon myself the responsibility of getting the armament on board *Dolphin*, and getting her into a position to defend the upper part of the yard."

[3] Senate Report No. 37, 26, Paulding's testimony.

[4] Ibid., 37, Pendergrast's testimony; Ibid., 92-93, Navy Department statement 14 November 1861; *Navy Register*, 1861, 102.

[5] *ORN*, I, 4, 293-96, Captain Wilkes to Secretary Welles, 22 April 1861. Figures for the required officers and men are taken from the Navy's estimates for moving the four ships out of Norfolk.

[6] That Paulding carried the four officers' orders for detachment is speculation, but the dating of their orders together with the date of his visit to Norfolk makes it likely that is what happened.

[7] *Navy Register*, 1861, 16-17.

[8] Senate Report No. 37, 50, McCauley's testimony. The commodore gave the impression of being somewhat peeved at how he had been treated by Secretary Toucey and possibly Welles, too.

[9] *ORN*, I, 4, 305, Commodore Paulding's narrative, 2 May 1861; Senate Report No. 37, 26, Paulding's testimony; Lull, *Gosport*, 53; Long, "Gosport Affair," 165.

[10] Senate Report No. 37, 26, 32, 49, Paulding's and McCauley's testimony; Rebecca Paulding Meade, *Life of Hiram Paulding, Rear-Admiral, U.S.N.* (New York: The Baker & Taylor Company, 1910), 238. Hereinafter cited as Meade, *Paulding*. Also see Still, *Selfridge*, 26-27.

[11] Senate Report No. 37, 26, Paulding's testimony; Lull, *Gosport*, 53; Still, *Selfridge*, 27.

[12] Meade, *Paulding*, 238.

[13] Senate Report No. 37, 26, Paulding's testimony.

[14] *ORN*, I, 4, 305, Commodore Paulding's narrative, 2 May 1861.

[15] Meade, *Paulding*, 238; Senate Report No. 37, 26, Paulding's testimony.

[16] Still, *Selfridge*, 27; Lull, *Gosport*, 53; *ORN*, I, 4, 277, Secretary Welles to Commodore McCauley, 16 April 1861.

[17] Senate Report No. 37, 29-30, Paulding's testimony.

[18] Ibid., 32-33, Paulding's testimony; Ibid., 43, Pendergrast's testimony.

[19] Ibid., 13, Introduction under "Captain Paulding."

[20] *ORN*, I, 4, 279, Secretary Welles to Commodore McCauley, 17 April 1861.

[21] Ibid., 280, Commodore Breese to Secretary Welles, 17 April 1861.

[22] Ibid., 280, Secretary Welles to Lieutenant Commanding Ronckendorff, 17 April 1861.

[23] Flanders, *Porter*, 37. Porter would be the naval constructor responsible for building the ironclad ram *Virginia*.

[24] *ORN*, I, 4, 280-81, Engineer-in-Chief Isherwood to Secretary Welles, 18 April 1861; Bennett, *Steam Navy*, 238.

CHAPTER 12 *Virginia Votes for Secession*

[1] Wise, *Henry Wise*, 277; Freehling, *Showdown*, 189.

[2] Wise, *Henry Wise,* 277.

[3] Imboden, *Battles and Leaders,* I, 112.

[4] Wise, *Henry Wise,* 277.

[5] Ibid., 278; Simpson, *A Good Southerner,* 250.

[6] Ibid.

[7] Wise, *Henry Wise,* 279-80.

[8] Ibid.

[9] *Journal of the Acts and Proceedings of a General Convention of the State of Virginia* (Richmond: Wyatt M. Elliott, Printer, 1861). Portions of Journal of Secret Session of the Convention, withheld from publication at its session ending May 1, 1861, 10-12. Hereinafter cited as *Journal of the Convention.*

[10] Freehling, *Showdown,* 196-97.

[11] Ibid., 198-99. Wise corrected Randolph's exaggerated quantity of gun powder.

[12] *Journal of the Convention,* Ordinances, 1, 6.

[13] Simpson, *A Good Southerner,* 229.

[14] Werlich, *Admiral of the Amazon,* 15-16; Dudley, *Going South,* 35-55.

[15] "Burning of Gosport Navy-Yard," *The New York Times,* 24 April 1861.

[16] Ibid.; *ORN,* I, 4, 248-50, Assistant Secretary Fox to Secretary Welles, 24 February 1865.

[17] Ibid., NHHC website, "DANFS," *Yankee* (1) Tug, 1861-1865, accessed November 2022.

[18] "Burning of Gosport Navy-Yard," *The New York Times,* 24 April 1861.

[19] Niven, *Welles,* 342.

[20] Welles, *Diary,* I, 46, 47.

[21] Ibid.

[22] Meade, *Paulding,* 238; *ORN,* I, 4, 305, Commodore Paulding's narrative, 2 May 1861.

[23] Madeleine Vinton Dahlgren, *Memoir of John A. Dahlgren, Rear-Admiral United States Navy* (Boston: James R. Osgood and Company, 1882), 330. Note that Dahlgren mistook the date the Convention went into secret session, Tuesday, for the day the ordinance of secession was passed, Wednesday 17 April.

[24] *ORN,* I, 4, 283, Secretary Welles to Commander Hunter, 18 April 1861.

[25] Ibid., 283, Secretary Welles to Commodore Breese, 18 April 1861.

[26] Ibid., 282, Secretary Welles to Lieutenant Woodhull, 18 April 1861; Ibid., Secretary Welles to Commodore Du Pont, 18 April 1861; Ibid., 283-84, Commodore Du Pont to Secretary Wells, 19 April 1861.

CHAPTER 13 *The Commodore's Decision*

[1] *ORN,* I, 4, 280-81, Engineer Isherwood to Secretary McCauley, 18 April 1861.

[2] Ibid.

[3] Bennett, *Steam Navy*, 238.

[4] *ORN*, I, 4, 300-02, Commander Alden to Secretary Welles, 30 November 1861.

[5] Ibid.

[6] NHHC website, "Officers List—who resigned at the outbreak of the Civil War," https://www.history.navy.mil/research/library/manuscripts/o/officers-list-who-resigned-at-the-outbreak-of-the-civil-war.html, accessed August 2021. Hereinafter cited as NHHC Website, "Officers List." Five of the yard's key commanders and lieutenants resigned on the 18th and others did so in the days that followed.

[7] *ORN*, I, 4, 300-02, Commander Alden to Secretary Welles, 30 November 1861.

[8] Ibid.

[9] Senate Report No. 37, 48-49, McCauley's testimony.

[10] *ORN*, I, 4, 300-02, Commander Alden to Secretary Welles, 30 November 1861.

[11] Welles, *Diary*, I, 44.

[12] *ORN*, I, 4, 280-82, Engineer-in-Chief Isherwood to Secretary Welles, 18 April 1861.

[13] Senate Report No. 37, 49, McCauley's testimony.

[14] Welles, *Diary*, I, 44-45.

[15] Bennett, *Steam Navy*, 238.

[16] Ibid., 240.

[17] Welles, *Diary*, I, 44.

[18] Wilkes, *Autobiography*, 762.

[19] *ORN*, I, 4, 293-96, Captain Wilkes to Flag Officer Paulding, 22 April 1861; Ibid., 300-02, Commander Alden to Secretary Welles, 30 November 1861.

[20] Bennett, *Steam Navy*, 237.

[21] *ORN*, I, 4, 300-02, Commander Alden to Secretary Welles, 30 November 1861.

[22] Senate Report No. 37, 49, McCauley's testimony.

[23] NARA, RG 45, M125, Roll 355, Captains' Letters, Captain Pendergrast to Secretary Dobbin, 12 March 1856; Ibid., RG 24, Deck Log of U.S.S. *Merrimack*, 9 March 1856; Kinnaman, *Merrimack*, 41.

[24] Bennett, *Steam Navy*, 241.

[25] Senate Report No. 37, 60, Doran's testimony.

CHAPTER 14 *Alden's Shocking Report*

[1] *ORN*, I, 4, 280-81, Engineer-in-Chief Isherwood to Secretary Welles, 18 April 1861.

[2] Ibid., 300-02, Commander Alden to Secretary Welles, 30 November 1861.

[3] Senate Report No. 37, 60, 79, Testimonies of Doran and Isherwood.

[4] Imboden, *Battles and Leaders*, I, 115-18.

[5] Welles, *Diary*, I, 44. Alden's report stated that he verbally reported to Welles on the 19th.

[6] Welles, *Civil War*.

[7] Niven, *Welles*, 577-78; Richard S. West, Jr., *Gideon Welles, Lincoln's Navy Department* (Indianapolis and New York: The Bobbs-Merrill Company, 1943), 193-94.

[8] Welles, *Civil War*, 34.

[9] Welles, *Diary*, I, 44-45.

[10] Ibid.; Welles, *Civil War*, 31.

[11] Bennett, *Steam Engineering*, 241.

[12] Welles, *Diary*, I, 45.

[13] "Southern Ports to be Blockaded" and "Startling from Baltimore," *The New York Times*, 20 April 1861.

[14] Welles, *Civil War*, 32; *ORN*, I, 4, 271, John Rodney to Secretary Welles, 18 April 1861.

[15] Welles, *Diary*, I, 45-46.

[16] Long, "Gosport Affair," 160.

[17] Welles, *Civil War*, 32-33.

[18] Ibid., 32.

[19] *ORN*, I, 4, 282, Secretary Welles to Commodore Paulding, 18 April 1861. This order may have been verbally issued on this date but there is reason to believe it was written on the following day, 19 April.

[20] Niven, *Welles*, 344.

[21] *ORN*, I, 4, 289, Commodore Paulding to Secretary Welles, 23 April 1861.

[22] Ibid., 305, Commodore Paulding's narrative, 2 May 1861.

[23] Meade, *Paulding*, 239.

[24] NHHC website, "DANFS," *Anacostia I* (Screw Steamer), accessed November 2022.

[25] *ORN*, I, 4, 284, Secretary Welles to Commodore Buchanan, 19 April 1861; Ibid., Secretary Welles to Commander Dahlgren, 19 April 1861.

[26] *ORN*, I, 4, 287-88, Commander Dahlgren to Secretary Welles, 20 April 1861.

[27] Meade, *Paulding*, 239.

[28] *ORN*, I, 4, 254-55, Abstract log of U.S.S. *Pawnee*, 9-19 April 1861.

[29] NHHC website, "DANFS," *Pawnee I* (Screw Sloop), accessed November 2022. Also see Donald L. Canney, *The Old Steam Navy, Volume One: Frigates, Sloops and Gunboats, 1815-1885* (Annapolis: Naval Institute Press, 1990), 84-87. Hereinafter cited as Canney, *Old Steam Navy I.*

[30] Meade, *Paulding*, 239.

[31] *ORN*, I, 4, 284, Secretary Welles to Colonel Harris, 19 April 1861.

[32] Ibid., 285, two orders, Secretary Welles to Commander Rowan, 19 April 1861.

[33] Paullin, *History*, 259.

[34] NARA, M330, Roll 9, various names. Orders for the officers' special or temporary duty were issued over two days, 18 April (seven) and 19 April (five including Wilkes). Alden was still under orders dated 11 April 1861.

[35] *ORN*, I, 4, 293-96, Captain Charles Wilkes to Flag Officer Paulding, 22 April 1861.

[36] Ibid.

[37] Meade, *Paulding*, 239.

[38] *ORN*, I, 4, 286, Commodore Breese to Secretary Welles, 19 April 1861.

[39] Ibid., 285-86, Commander Hunter to Secretary Welles, 19 April 1861.

[40] Niven, *Welles*, 344. Welles' other mistake, according to Niven, was giving command of the relief expedition to Paulding.

[41] Scharf, *Confederate Navy*, 138.

[42] Senate Report No. 37, 51, 68, Wise and Russell's testimony.

CHAPTER 15 *Norfolk Showdown*

[1] Tucker, *Imboden*, 30-31; *ORN*, I, 4, 306-09, General Taliaferro to Governor Letcher, 23 April 1861.

[2] *ORN*, I, 4, 306-09, General Taliaferro to Governor Letcher, 23 April 1861; Porter, *A Record*, 11.

[3] *Lamb's Biographical Dictionary of the United States*, John Howard Brown, Ed. (Boston: Federal Book Company of Boston, 1903), Vol. VII, 278-79. Hereinafter cited as *Lamb's*, Taliaferro.

[4] *ORN*, I, 4, 306-09, General Taliaferro to Governor Letcher, 23 April 1861.

[5] Ibid.

[6] Ibid.

[7] NHHC website, "Officers List." Also see Appendix 2, List of Officers at the Gosport Navy Yard, for a graphic representation of the dates of the Southern officers' resignations.

[8] Senate Report No. 37, 49, McCauley's testimony; Ibid., 92, Statement of the number of marines … at the Norfolk Navy Yard, dated 14 November 1861.

[9] Porter, *A Record*, 11; William Wirt Henry and Ainsworth R. Spofford, *Eminent and Representative Men of Virginia and the District of Columbia of the Nineteenth Century* (Madison, WI: Brant & Fuller, 1893), 508-13, William Mahone.

[10] Lull, *Gosport*, 52.

[11] Senate Report No. 37, 19, opening statement.

[12] *ORN*, I, 4, 306-09, General Taliaferro to Governor Letcher, 23 April 1861.

[13] Senate Report No. 37, 47, McCauley's testimony.

[14] Ibid., 38, Pendergrast's testimony; Ibid., 76, Clements' testimony.

[15] Lull, *Gosport*, 54; *ORN*, I, 4, 306, G. T. Sinclair to Secretary Mallory, 22 April 1861.

[16] *ORN*, I, 4, 306-09, General Taliaferro to Governor Letcher, 23 April 1861.

[17] Still, *Selfridge*, 28.

[18] *ORN*, I, 4, 288-89, Commodore McCauley to Secretary Welles, 25 April 1861.

[19] NHHC Website, "Officers," under Robert R. Pegram and Thomas O. Selfridge, Jr.

[20] Still, *Selfridge*, 28.

[21] Senate Report No. 37, 47, McCauley's testimony

[22] Still, *Selfridge*, 29.

[23] Senate Report No. 37, 73, Clements' testimony.

CHAPTER 16 *McCauley's Scuttling*

[1] Senate Report No. 37, 71, Brownley's testimony.

[2] Col. William H. Stewart, *History of Norfolk County, Virginia, and Representative Citizens.* (Chicago: Biographical Publishing Company, 1902), 448. Pages 448-453 contain a facsimile of "An Account of the Evacuation of the Navy Yard, First by the Federals in April, 1861, and Second by the Confederates in May 1862," by William H. Peters, Navy Agent, C.S. Hereinafter cited as Stewart, *History.*

[3] Senate Report No. 37, 38, Pendergrast's testimony. The commodore stated that the attempted seizure of *Yankee* occurred "on the 19th" which may have been the case. However, I have cast the narrative to place it on Saturday the 20th, in conformance with the story of eyewitness Daniel O'Connor, whose account is referenced in the note below. Finally, an article about *Yankee* appearing in *The New York Times,* also referenced in the note below, puts the event on "Friday the 18th" serving only to further confuse the true date of her attempted seizure.

[4] "Burning of Gosport Navy-Yard," *The New York Times,* 24 April 1861.

[5] John W. Connor, "A letter from Daniel O'Connor, USMC, reports: The North's Fiasco at Norfolk Navy Yard," *Military Images,* Vol. 20, No. 3 (November-December, 1998): 29-31. Hereinafter cited as Connor, "A letter from Daniel O'Connor."

[6] Senate Report No. 37, 38, Pendergrast's testimony.

[7] Ibid., 67, Edelin's testimony.

[8] Porter, Admiral David D., *Naval History of the Civil War* (New York: The Sherman Publishing Company, 1886), 28. Hereinafter cited as Porter, *Naval History.*

[9] John V. Quarstein, "The Capture of Gosport Navy Yard," *Civil War Navy—the Magazine,* Vol. 5, Issue 4 (Spring 2018): 20.

[10] Connor, "A letter from Daniel O'Connor." Punctuation and capitalizations have been added to clarify the intent and improve the flow.

[11] Senate Report No. 37, 90-92, No. 4, List of officers on duty at the Norfolk navy yard ...

[12] Ibid., 47, McCauley's testimony.

[13] Ibid., 41, Pendergrast's testimony.

[14] Ibid., 47, McCauley's testimony.

[15] Ibid., 43, Pendergrast's testimony.

[16] Lull, *Gosport,* 62.

[17] Canney, *Sailing Warships,* 192.

[18] Senate Report No. 37, 43, Pendergrast's testimony.

[19] Ibid., 47-48, McCauley's testimony.

[20] Ibid.

[21] *ORN,* I, 4, 288-89, Commodore McCauley to Secretary Welles, 25 April 1861.

[22] Porter, *A Record,* 13.

[23] Kinnaman, *Merrimack,* 242. These valves were used to routinely "blow down" excessively salty boiler water and eject it outside the hull.

[24] Still, *Selfridge,* 30-31.

[25] Ibid.; Porter, *A Record,* 13.

[26] *The Oxford Companion to Ships and the Sea*, Peter Kemp, Ed. (Oxford University Press, 1979), p. 778, Sheer Legs or Sheers. Contemporary U.S. Navy usage calls these devices "shears," whereas Oxford spells the word "sheers."

[27] Meade, *Paulding*, 240.

[28] Porter, *A Record*, 13.

[29] Stewart, *History*, 448.

[30] Porter, *A Record*, 14; John Quarstein, *The CSS Virginia, Sink Before Surrender* (Charleston and London: The History Press, 2012), 39. Hereinafter cited as Quarstein, *CSS Virginia*.

[31] Welles, *Diary*, I, 52.

CHAPTER 17 *The Die is Cast*

[1] *ORN*, I, 4, 292, Abstract of log of *Pawnee*, 20 April 1861.

[2] Quarstein, *CSS Virginia*, 41.

[3] *ORN*, I, 4, 292, Abstract of log of *Pawnee*, 20 April 1861; Senate Report No. 37, 51, Wise's testimony; Wilkes, *Autobiography*, 760.

[4] *ORN*, I, 4, 292, Abstract of log of *Pawnee*, 20 April 1861; Ibid., 293-96, Captain Wilkes to Flag Officer Paulding, 22 April 1861. *Pawnee*'s trouble-free passage was due to the skills of her master, Captain Mitchell, "an old and experienced pilot." See Senate Report No. 37, 42, Pendergrast's testimony. There is reason to believe Mitchell may have been a Bay Line captain.

[5] Still, *Selfridge*, 31-32.

[6] Wilkes, *Autobiography*, 760, 763. Wilkes was blunt about *Pawnee*'s narrow escape from disaster due her "omission of proper precautions," the making a signal of recognition.

[7] *ORN*, I, 4, 293-96, Captain Wilkes to Flag Officer Paulding, 22 April 1861.

[8] Wilkes, *Autobiography*, 760. In this same description, Wilkes tells of how Edelin earnestly begged that a building be spared destruction because it was home to a hen sitting on her eggs.

[9] Ibid.

[10] *ORN*, I, 4, 296-98, Captain Wright to Lieutenant Colonel Townsend, 26 April 1861.

[11] Senate Report No. 37, 48, McCauley's testimony; Ibid., 42, Pendergrast's testimony.

[12] *ORN*, I, 4, 289-91, Commodore Paulding to Secretary Welles, 23 April 1861.

[13] *Oxford Essential Quotations*, 4th Ed, Susan Ratcliffe, Ed. (Oxford University Press, 2016). The edited phrase was first quoted by Helmuth von Moltke in 1880.

[14] *ORN*, I, 4, 305, Commodore Paulding's narrative, 2 May 1861; Ibid., 293-96, Captain Wilkes to Flag Officer Paulding, 22 April 1861; Senate Report No. 37, 51, Wise's testimony.

[15] Senate Report No. 37, 30, Paulding's testimony; Porter, *A Record*, 17.

[16] *ORN*, I, 4, 289-91, Commodore Paulding to Secretary Welles, 23 April 1861; Porter, *A Record*, 17.

[17] Senate Report No. 37, 36, Paulding's testimony.

[18] *ORN*, I, 4, 289-91, Commodore Paulding to Secretary Welles, 23 April 1861.

[19] Wilkes, *Autobiography*, 763; Senate Report No. 37, 41, Pendergrast's testimony.

[20] *ORN*, I, 4, 292, 91, Abstract Log of *Pawnee*; Senate Report No. 37, 32, Paulding's testimony.

[21] Porter, *A Record*, 17.

[22] Senate Report No. 37, 27-32, Paulding's testimony; *ORN*, I, 4, 305, Commodore Paulding's narrative, 2 May 1861.

[23] Wilkes, *Autobiography*, 761-62.

[24] Senate Report No. 37, 36, Paulding's testimony.

[25] *ORN*, I, 4, 282, Secretary Welles to Commodore Paulding, 18 April 1861.

[26] Philip Van Doren Stern, *The Confederate Navy, A Pictorial History* (Garden City: Doubleday & Company, Inc., 1962), 39.

[27] Senate Report No. 37, 27, Paulding's testimony; *ORN*, I, 4, 305, Commodore Paulding's narrative, 2 May 1861.

[28] Senate Report No. 37, 51, Wise's testimony.

[29] Still, *Selfridge*, 32-33.

[30] Wilkes, *Autobiography*, 762, 764-65.

CHAPTER 18 *Paulding's Destruction*

[1] *ORN*, 1, 4, 293-96, Captain Wilkes to Flag Officer Paulding, 22 April 1861; Wilkes, *Autobiography*, 762; Lull, *Gosport*, 56.

[2] Senate Report No. 37, 51-52, Wise's testimony.

[3] *ORN*, 1, 4, 293-96, Captain Wilkes to Flag Officer Paulding, 22 April 1861; Lull, *Gosport*, 36, the graving dock was valued at $974,356.65. Also see Charles B. Stuart, *The Naval Dry Docks of the United States* (New York: Charles B. Norton, 1852), 72-73.

[4] *ORN*, I, 4, 296-98, Captain Wright to Lieutenant Colonel Townsend, 26 April 1861.

[5] Porter, *A Record*, 15.

[6] Senate Report No. 37, 52, Wise's testimony; *ORN*, 1, 4, 293-96, Captain Wilkes to Flag Officer Paulding, 22 April 1861.

[7] Senate Report No. 37, 52, Wise's testimony.

[8] Ibid.; *ORN*, 1, 4, 293-96, Captain Wilkes to Flag Officer Paulding, 22 April 1861.

[9] Lull, *Gosport*, 44-45; Porter, *A Record*, 14; *ORN*, 1, 4, 293-96, Captain Wilkes to Flag Officer Paulding, 22 April 1861.

[10] *ORN*, 1, 4, 293-96, Captain Wilkes to Flag Officer Paulding, 22 April 1861; Senate Report No. 37, 52, Wise's testimony.

[11] *ORN*, 1, 4, 293-96, Captain Wilkes to Flag Officer Paulding, 22 April 1861.

[12] *ORN*, I, 4, 300-02, Commander Alden to Secretary Welles, 30 November 1861.

[13] Wilkes, *Autobiography*, 761.

[14] *ORN*, 1, 4, 293-96, Captain Wilkes to Flag Officer Paulding, 22 April 1861. In this same report, Wilkes identified the parties' rendezvous point as at the end of the "eastern ship house" (Ship House B) and the "lower ship house" (Ship House A). Other clues convince the author that in both cases Wilkes meant in front of Ship House A. He also implied Lieutenant Wise was supplied with a boat from *Cumberland*; Wise was already afloat in one of *Pawnee*'s boats.

[15] Senate Report No. 37, 49, Paulding's testimony; Wilkes, *Autobiography*, 761; *ORN*, I, 4, 292, 91, Abstract Log of *Pawnee*.

[16] *ORN*, 1, 4, 293-96, Captain Wilkes to Flag Officer Paulding, 22 April 1861.

[17] Ibid. Close examination of contemporary accounts reveals some of the officers reported times for the same event differing by as much as two hours, e.g., Wise reported seeing the commodore's signal at 2:20 a.m.; Senate Report No. 37, 52, Wise's testimony. The account provided herein adheres to times noted in the abstract log of *Pawnee*.

[18] Senate Report No. 37, 52, Wise's testimony.

[19] *ORN*, 1, 4, 293-96, Captain Wilkes to Flag Officer Paulding, 22 April 1861.

[20] *ORN*, I, 4, 292-91, Abstract Log of *Pawnee*; Senate Report No. 37, 53, Wise's testimony.

[21] Porter, *A Record*, 16.

[22] Ibid.; *ORN*, 1, 4, 293-96, Captain Wilkes to Flag Officer Paulding, 22 April 1861.

[23] *ORN*, I, 4, 292-91, Abstract Log of *Pawnee*; Senate Report No. 37, 53, Wise's testimony.

[24] *ORN*, I, 4, 292-91, Abstract Log of *Pawnee*; Ibid., 287, Commodore Du Pont to Secretary Welles, 20 April 1861.

[25] *ORN*, I, 4, 289-91, Commodore Paulding to Secretary Welles, 23 April 1861; Ibid., 288-89, Commodore McCauley to Secretary Welles, 25 April 1861; Still, *Selfridge*, 34-35.

[26] Paul H. Silverstone, *Warships of the Civil War* (Annapolis: Naval Institute Press, 1989), 27, 127-28.

[27] Kinnaman, *Merrimack*, 58; Senate Report No. 37, 49-50, McCauley's testimony.

[28] *ORN*, I, 4, 298, Commodore Paulding to Flag Officer Pendergrast, 22 April 1861; Ibid., 298, Commodore Paulding to Captain Rowan, 22 April 1861; Ibid., 299, twice from Commodore Paulding to Flag Officer Pendergrast, 22 April 1861.

[29] Ibid., 289-91, Commodore Paulding to Secretary Welles, 23 April 1861.

CHAPTER 19 *The Aftermath*

[1] Porter, *A Record*, 17.

[2] Ibid., 14-16; Stewart, *History*, 449. Could it be that the compassionate seaman was John Reynolds?

[3] *ORN*, I, 4, 306, G. T. Sinclair to Secretary Mallory, 22 April 1861.

[4] Senate Report No. 37, 70, Brownley's testimony.

[5] Stewart, *History*, 450.

[6] Senate Report No. 37, 70, Brownley's testimony.

[7] Ibid., 58, Toucey's testimony. Also see *ORN*, I, 4, 302-03, Toucey's testimony.

[8] Senate Report No. 37, 70, Brownley's testimony; *ORN*, I, 4, 306, G. T. Sinclair to Secretary Mallory, 22 April 1861.

[9] Porter, *Naval History*, 32.

[10] *Message of the President of the United States to the Two Houses of Congress at the Commencement of the Second Session of the Thirty-Seventh Congress*, Vol. III. (Washington: Government Printing Office, 1861), Annual Report of the Secretary of the Navy, 2 December 1861, 22. Hereinafter cited ARSN 1861.

[11] Senate Report No. 37, 90, No. 3, Summary Statement of the public property ...

[12] "The War News," "Burning of Gosport Navy-Yard," *The New York Times*, 24 April 1861.

[13] Paullin, *History*, 274.

[14] Coletta, *Secretaries*, I, 326.

[15] *ORN*, 1, 4, 293-96, Captain Wilkes to Flag Officer Paulding, 22 April 1861.

[16] Ibid., 289-91, Commodore Paulding to Secretary Welles, 23 April 1861; Ibid., 298, Captain Paulding to Secretary Welles, 24 April 1861.

[17] Ibid., 305, Commodore Paulding's narrative, 2 May 1861.

[18] Ibid., 53-55, Naval General Order, Secretary Welles, 24 April 1861.

[19] Ibid., 288-89, Commodore McCauley to Secretary Welles, 25 April 1861.

[20] Ibid., 296-98, Captain Wright to Lieutenant Colonel Townsend, 26 April 1861.

[21] Ibid., 300-02, Commander Alden to Secretary Welles, 30 November 1861.

[22] *ORN*, II, 2, 67-69, S. R. Mallory to C. M. Conrad, 10 May 1861; William C. Davis, *Duel Between the First Ironclads* (Baton Rouge and London: Louisiana State University Press, 1975), 8. Hereinafter cited as Davis, *Duel*.

[23] Davis, *Duel*, 12, 26.

[24] *ORN*, II, 2, 70-72, Secretary Mallory to Lieutenant North, 17 May 1861.

[25] Ibid., 76-79, Secretary Mallory to the President, 18 July 1861.

[26] Davis, *Duel*, 12.

[27] Ibid., 32.

[28] Ibid., 35-36.

[29] *ORN*, I, 6, 776-77, Secretary Mallory to Flag Officer Buchanan, 24 February 1862.

[30] Davis, *Duel*, 80.

[31] Ibid., 79; Still, *Selfridge*, 43.

[32] Davis, *Duel*, 83-89.

[33] Still, *Selfridge*, 46-54.

[34] Ibid., 55-56.

[35] George F. Amadon, *Rise of the Ironclads* (Missoula, MT: Pictorial Histories Publishing Company, 1990), 43.

CHAPTER 20 *The Accounting*

[1] *ORN*, I, 4, 289-91, Commodore Paulding to Secretary Welles, 23 April 1861.

[2] NARA, M330, Roll 9, Captains—Active List: 17 Hiram Paulding, 30 Garrett J. Pendergrast, 8 Charles S. McCauley "May 13/61 Det'd 20th & wait orders."

[3] Senate Report No. 37, 1, Report opening statement.

[4] Welles, *Diary*, I, 53.

[5] Bennett, *Steam Engineering*, 236.

[6] *ORN*, I, 4, 305, Commodore Paulding's narrative, 2 May 1861; "The Destruction of Norfolk Navy Yard," *The Evening Post*, New York, 28 April 1862.

[7] *ORN*, I, 4, 18, Commodore Armstrong to Secretary Toucey, 26 January 1861.

[8] Ivan Musicant, "The Fires of Norfolk," *American Heritage*, Vol. 41, Issue 2 (March 1990). Musicant confirms that "The Navy convened neither general courts-martial nor a board of inquiry into the facts of the Norfolk debacle."

[9] Welles, *Diary*, I, 43.

[10] McCauley Obituary, Boston *Pilot*.

[11] NARA, M330, Roll 9, Captains—Active List: 17 Hiram Paulding, 30 Garrett J. Pendergrast, 7 Francis H. Gregory; James Phinney Baxter, III, *The Introduction of the Ironclad Warship* (Annapolis: Naval Institute Press, 2001), 247-48; *The Blockade, Runners and Raiders*, Gerald Simons, Ed. (Alexandria, VA: Time-Life Books, Inc., 1983), 24; Kinnaman, *Lenthall*, 243, 248.

[12] *Civil War Naval Chronology, 1861-1865* (Washington: Naval History Division, Navy Department, 1971), I-31-34.

[13] Welles, *Diary*, I, 48-49, 53.

[14] Senate Report No. 37, 71, Brownley's testimony.

[15] Ibid., 50, McCauley's testimony.

[16] Ibid., 20-21, Report summary.

[17] Paullin, *History*, 273.

[18] Ibid., 64, Edelin's testimony.

[19] ARSN 1861, 21.

[20] Du Pont, *Civil War Letters*, II, 35, journal letter to Mrs. Du Pont, 5 May 1862.

[21] Kinnaman, *Merrimack*, 283.

[22] McCauley Obituary, Boston *Pilot*.

[23] "A Hero Gone to His Rest, Death of Rear-Admiral Paulding," *The New York Times*, 21 October 1878.

[24] Welles, *Civil War*, 35; Coletta, *Secretaries*, I, 321-357.

[25] Hamersly, *Records*, 36.

[26] Sloan, *Isherwood*, 27-243.

[27] Still, *Selfridge*, vii-xxxvi.

[28] Frank J. Merli, *Great Britain and the Confederate Navy, 1861-1865* (Bloomington and London: Indiana University Press, 1970), 120-26; Werlich, *Admiral of the Amazon*, 23-240; *Lamb's*, Page.

[29] Davis, *Dual*, 27, 35; Simpson, *A Good Southerner*, 252-315.

[30] Butt, *Norfolk*, 13.

[31] NARA, M330, Roll 9, Commanders—Active List, 191 John W. Livingston. "May 20/62, to command Naval Station, Norfolk."

APPENDIX I *Annotated Calendar, March and April 1861*

[1] *ORN*, I, 4, xvii, Calendar 1861.

[2] Potter, *Impending Crisis*, 494-99.

APPENDIX 2 *List of Officers at the Gosport Navy Yard*

[1] *Navy Register*, 1861, 100; adjusted for corrections noted in Senate Report No. 37, 91-92. Officers on special duty or who were in transit or temporary visitors are not shown on the first table A2.1, but are included in the more comprehensive tables A2.2 and A2.3.

APPENDIX 3 *Secretary Welles' Orders of 16 April 1861*

[1] *ORN*, I, 4, 277-78, Secretary Welles to Commodore McCauley, 16 April 1861.

[2] Ibid., 278, Secretary Welles to Commodore Pendergrast, 16 April 1861.

APPENDIX 4 *Fate of Warships at the Gosport Navy Yard*

[1] *Navy Register*, 1861, 95-97.

[2] Ibid.

[3] Porter, *A Record*, 12.

[4] Senate Report No. 37, 38, 52, 63, 70, Pendergrast, Wise, Doran and Brownley's testimonies, respectively; *ORN*, I, 4, 306, G. T. Sinclair to S. R. Mallory, 22 April 1861.

APPENDIX 5 *Destination of the Guns Seized from the Gosport Navy Yard*

[1] Porter, *Naval History*, 32.

[2] Senate Report No. 37, 45, 30, 87-88, McCauley and Paulding's testimonies, Appendix, No. 2, Harwood's statement, respectively.

[3] Stewart, *History*, 450-52.

[4] Senate Report No. 37, 75, Clements' testimony.

[5] Porter, *Naval History*, 32.

[6] Spencer Tucker, *Arming the Fleet, U.S. Navy Ordnance in the Muzzle-Loading Era* (Annapolis: Naval Institute Press, 1989), 199-200.

[7] Raphael Semmes, *Memoirs of Service Afloat During the War Between the States* (Baltimore: Kelly, Piet & Co., 1869), 99-120.

[8] Stewart, *History*, 450-52. A total of 670 guns were listed by Peters.

APPENDIX 6 *The Potential of Conversions to a Steam Frigate*

[1] NARA, RG 45, M518, Roll 14, John Lenthall to Secretary Toucey, 18 August 1857.

[2] "NAVAL," Richmond *Daily Dispatch*, 15 April 1860; "NEWS OF THE DAY," *Alexandria Gazette*, 15 April 1860.

[3] Senate Report No. 37, 58-59, Toucey's testimony.

***

BIBLIOGRAPHY

Government Documents and Publications

National Archives and Records Administration (NARA), Washington, D.C.:

- Record Group 19, Records of the Bureau of Ships, Bureau of Steam Engineering, Alphabetical File—Merrimack.
- Record Group 24, Records of the Bureau of Naval Personnel, Abstracts of Service Records of Naval Officers, 1789-1893, M330.
- Record Group 24, Records of the Bureau of Naval Personnel, Deck Logs, USS *Merrimack*.
- Record Group 36, Records of the U.S. Customs Service, Proofs of Citizenship used to apply for Seamen's Certificates for the Port of Philadelphia, Pennsylvania, 1792-1861, M1880.
- Record Group 45, Naval Records Collection of the Office of Naval Records and Library, Letters Received by the Secretary of the Navy from Captains, 1805-1861, M125.
- Record Group 45, Naval Records Collection of the Office of Naval Records and Library, Records of the Office of the Secretary of the Navy, Letters Received by the Secretary from Chiefs of Navy Bureaus, 1842-1885, M518.

State of Virginia:

- *Journal of the Acts and Proceedings of a General Convention of the State of Virginia.* Richmond: Wyatt M. Elliott, Printer, 1861.

U. S. Department of the Navy:

- *Autobiography of Rear Admiral Charles Wilkes, U.S. Navy 1798-1877.* William James Morgan, et al., Eds. Washington: Naval History Division, Department of the Navy, 1978.
- *Civil War Naval Chronology, 1861-1865.* Washington: Naval History Division, Navy Department, 1971.
- *Dictionary of American Naval Fighting Ships* (DANFS). Washington, D.C.: Office of the Chief of Naval Operations, Naval History Division, 1969.
- Lull, Edward P. *History of the United States Navy-Yard at Gosport, Virginia (near Norfolk).* Washington: Government Printing Office, 1874.

- *Official Records of the Union and Confederate Navies in the War of the Rebellion (ORN).* 30 volumes and index. Washington, D.C.: U.S. Government Printing Office, 1894-1927.
- *Register of Commissioned and Warrant Officers of the Navy of the United States, including Officers of the Marine Corps, and Others* (*Navy Register*). Washington: Government Printing Office, various years.

U.S. House of Representatives:

- 37th Congress, 2d Session, Executive Document. *Message of the President of the United States to the Two Houses of Congress at the Commencement of the Second Session of the Thirty-Seventh Congress*, Vol. III. Washington: Government Printing Office, 1861.

U. S. Senate:

- 37th Congress, 2d Session, Senate Report No. 37. *Report of the Select Committee appointed to inquire into the circumstances attending the surrender of the Navy Yard at Pensacola, and the Destruction of the Property of the United States at the Navy Yard at Norfolk, and at the Armory at Harper's Ferry.* Washington: Government Printing Office, 1862.

Published Memoirs and Papers

Dahlgren, Madeleine Vinton. *Memoir of John A. Dahlgren, Rear-Admiral United States Navy.* Boston: James R. Osgood and Company, 1882.

Du Pont, Samuel Francis. *Samuel Francis Du Pont, A Selection from his Civil War Letters.* John D. Hayes, Ed. 3 Vols. Ithaca: Cornell University Press, 1969.

Russell, William Howard. *My Diary, North and South.* Boston: T. O. H. P. Burnham, 1863.

Semmes, Raphael. *Memoirs of Service Afloat During the War Between the States.* Baltimore: Kelly, Piet & Co., 1869.

Welles, Gideon. *Diary of Gideon Welles, Secretary of the Navy under Lincoln and Johnson.* 3 Vols. Boston and New York: Houghton, Mifflin Company, 1911.

Welles, Gideon. *Civil War and Reconstruction.* Albert Mordell, Ed. New York: TWayne Publishers, 1959.

Unpublished Papers Collections

U.S. Naval Academy Museum, The Ives Collection (John Lenthall papers).

Biographical Dictionaries and Encyclopedias

Appleton's Cyclopedia of American Biography. James Grant Wilson and John Fiske, Eds. New York: D. Appleton and Company, 1887-1889.

Lamb's Biographical Dictionary of the United States. John Howard Brown, Ed. Boston: Federal Book Company of Boston, 1903.

Oxford Essential Quotations, 4th Edition. Susan Ratcliffe, Ed. Oxford University Press, 2016.

The Oxford Companion to Ships and the Sea. Peter Kemp, Ed. Oxford University Press, 1979.

Contemporary Newspapers and Serials

Alexandria Gazette

Daily Dispatch

Harper's Weekly

Indiana American

The Congressional Globe

The Evening Post

The New York Times

The Pilot

Supporting Sources, Books

Amadon, George F. *Rise of the Ironclads.* Missoula, MT: Pictorial Histories Publishing Company, 1990.

Baxter, James Finney, III. *The Introduction of the Ironclad Warship.* Annapolis: Naval Institute Press, 2001.

Bennett, Frank M. *The Steam Navy of the United States.* Pittsburgh, PA: W. T. Nicholson, 1896.

Berube, Claude. *On Wide Seas, The US Navy in the Jacksonian Era.* Tuscaloosa: The University of Alabama Press, 2021.

Berube, Claude and John Rodgaard. *A Call to the Sea, Captain Charles Stewart of the USS Constitution.* Washington, D.C.: Potomac Books, Inc, 2006.

Bradford, James E., Ed. *Command Under Sail, Makers of the American Naval Tradition 1775-1850.* Annapolis: Naval Institute Press, 1985.

Brown, Alexander Crosby. *Steam Packets on the Chesapeake, A History of the Old Bay Line Since 1840.* Centreville, MD: Tidewater Publishers, 1961.

Butt, Marshall W. *A Brief History of Norfolk Naval Shipyard, Portsmouth, Virginia.* Portsmouth: Public Information Office, 1956.

————. *Portsmouth Under Four Flags, 1752-1970.* Portsmouth: Messenger Printing Co., Inc., 1971.

Canney, Donald L. *Lincoln's Navy, The Ships, Men and Organization, 1861-1865.* Annapolis: Naval Institute Press, 1998.

————. *Sailing Warships of the U.S. Navy.* Annapolis: Naval Institute Press, 2001.

————. *The Old Steam Navy, Volume I, Frigates, Sloops and Gunboats, 1815—1885.* Annapolis: Naval Institute Press, 1990.

Coletta, Paolo E., Ed. *American Secretaries of the Navy,* 1775-1972. 2 Vols. Annapolis: Naval Institute Press, 1980.

Davis, William C. *Duel Between the First Ironclads.* Baton Rouge and London: Louisiana State University Press, 1975.

Dudley, William S. *Going South: U.S. Navy Officer Resignations & Dismissals On The Eve Of The Civil War.* Washington, D.C.: Naval Historical Foundation, 1981.

Flanders, Alan B. *John L. Porter, Naval Constructor of Destiny.* Whitestone, VA: Brandylane Publishers, Inc., 2000.

Freehling, William W. and Craig M. Simpson, Eds. *Showdown in Virginia, The 1861 Convention and the Fate of the Union.* Charlottesville and London: University of Virginia Press, 2010.

Hamersly, Lewis Randolph. *The Records of Living Officers of the U.S. Navy and Marine Corps.* Philadelphia: J. B. Lippincott & Co., 1870.

Henry, William Wirt and Ainsworth R. Spofford. *Eminent and Representative Men of Virginia and the District of Columbia of the Nineteenth Century.* Madison, Wis: Brant & Fuller, 1893.

Hoogenboom, Ari. *Gustavus Vasa Fox of the Union Navy.* Baltimore: The Johns Hopkins Press, 2008.

Johnson, Robert Irwin. *Thence Around Cape Horn.* Annapolis: United States Naval Institute, 1963.

Johnson, Robert Underwood and Clarence Clough Buel, Eds. *Battles and Leaders of the Civil War.* New York: Century, Co., 1887.

Kinnaman, Stephen Chapin. *Captain Bulloch, The Life of James Dunwoody Bulloch, Naval Agent of the Confederacy.* Indianapolis: Dog Ear Publishing, 2013.

————. *Merrimack, The Biography of a Steam Frigate.* Wilmington and Malaga: Vernon Press, 2019.

————. *John Lenthall, The Life of a Naval Constructor.* Wilmington and Malaga: Vernon Press, 2022.

Klein, Maury. *Days of Defiance, Sumter, Secession, and the Coming of the Civil War.* New York: Alfred A. Knopf, 1997.

Long, David F. *Gold Braid and Foreign Relations, Diplomatic Activities of U.S. Naval Officers 1798-1883.* Annapolis: Naval Institute Press, 1988.

Lossing, Benson J. *Lossing's Pictorial Field Book of the War of 1812.* 2 Vols. New York: Harper & Brothers, 1868.

Meade, Rebecca Paulding. *Life of Hiram Paulding, Rear-Admiral, U.S.N.* New York: The Baker & Taylor Company, 1910.

Merli, Frank J. *Great Britain and the Confederate Navy, 1861-1865.* Bloomington and London: Indiana University Press, 1970.

Michael, Chris. *Lelia.* Birkenhead: Countryvise Limited, 2004.

Niven, John. *Gideon Welles, Lincoln's Secretary of the Navy.* Baton Rouge and London: Louisiana State University Press, 1973.

Paullin, Charles Oscar. *Paullin's History of Naval Administration 1775-1911.* Annapolis: U.S. Naval Institute, 1968.

Peck, Taylor. *Round-Shots to Rockets, The History of the Washington Navy Yard and the United States Naval Gun Factory.* Annapolis: United States Naval Institute, 1949.

Porter, Admiral David D. *Naval History of the Civil War.* New York: The Sherman Publishing Company, 1886.

Porter, John W. H. *A Record of Events in Norfolk County, Virginia, from April 19th, 1861, to May 10th, 1862 ….* Portsmouth, VA: W. A. Fiske, Printer and Bookbinder, 1892.

Potter, David M. *The Impending Crisis, 1848-1861.* New York: Harper & Row, Publishers, Inc., 1976.

Quarstein, John V. *The Battle of the Ironclads.* Charleston, etc.: Arcadia Publishing, 1999.

———.*The CSS Virginia, Sink Before Surrender.* Charleston and London: The History Press, 2012.

Quarstein, John V. and Dennis Mroczkowski. *Fortress Monroe, The Key to the South.* Charleston, etc.: Arcadia Publishing, 2000.

Scharf, J. Thomas. *History of the Confederate States Navy, from its Organization to the Surrender of its Last Vessel.* New York: Rogers & Sherwood, 1887.

Schroeder, John H. *Shaping a Maritime Empire, The Commercial and Diplomatic Role of the American Navy, 1829-1861.* Westport and London: Greenwood Press.

Silverstone, Paul H. *The Sailing Navy, 1775-1854.* Annapolis: Naval Institute Press, 2001.

———. *Warships of the Civil War Navies.* Annapolis: Naval Institute Press, 1989.

Simons, Gerald, Ed. *The Blockade, Runners and Raiders.* Alexandria, VA: Time-Life Books, Inc., 1983.

Simpson, Craig M. *A Good Southerner, The Life of Henry A. Wise of Virginia.* Chapel Hill and London: The University of North Carolina Press, 1985.

Sloan, Edward W., III. *Benjamin Franklin Isherwood, Naval Engineer.* Annapolis: United States Naval Institute, 1965.

Stern, Philip Van Doren. *The Confederate Navy, A Pictorial History*. Garden City: Doubleday & Company, Inc., 1962.

Stewart, Col. William H. *History of Norfolk County, Virginia, and Representative Citizens.* Chicago: Biographical Publishing Company, 1902.

Still, William N., Jr., Ed. *What Finer Tradition, The Memoirs of Thomas O. Selfridge, Jr., Rear-Admiral, U.S.N.* Columbia: University of South Carolina Press, 1987.

Stuart, Charles B. *The Naval Dry Docks of the United States.* New York: Charles B. Norton, 1852.

Tucker, Spencer. *Arming the Fleet, U.S. Navy Ordnance in the Muzzle-Loading Era.* Annapolis: Naval Institute Press, 1989.

————. *Brigadier General John D. Imboden, Confederate Commander in the Shenandoah.* Lexington: The University Press of Kentucky, 2003.

Tucker, Spencer and Frank T. Reuter. *Injured Honor, the Chesapeake-Leopard Affair, June 22, 1807.* Annapolis: Naval Institute Press, 1996.

Valle, James E. *Rocks and Shoals, Naval Discipline in the Age of Fighting Sail.* Annapolis: Naval Institute Press, 1980.

Ward, Geoffrey C. with Ken Burns and Ric Burns. *The Civil War.* New York: Alfred A. Knopf, Inc., 1990.

Werlich, David P. *Admiral of the Amazon: John Randolph Tucker, His Confederate Colleagues and Peru.* Charlottesville and London: University of Virginia Press, 1990.

West, Richard S., Jr. *Gideon Welles, Lincoln's Navy Department.* Indianapolis and New York: The Bobbs-Merrill Company, 1943.

Wise, Barton H. *The Life of Henry A. Wise of Virginia.* New York: The Macmillan Company, 1899.

Supporting Sources, Articles

Connor, John W. "A letter from Daniel O'Connor, USMC, reports: The North's Fiasco at Norfolk Navy Yard." *Military Images,* Vol. 20, No. 3 (November-December, 1998): 29-31.

Long, John Sherman. "The Gosport Affair, 1861." *The Journal of Southern History,* Vol. 23, No. 2 (May 1957): 155-172.

Musicant, Ivan. "The Fires of Norfolk." *American Heritage,* Vol. 41, Issue 2 (March 1990).

Nuermberger, Ruth Ketring. "Asbury Dickins (1780-1861): A Career in Public Service." *North Carolina Historical Review,* Vol. XXIV, No. 3 (July 1947): 281-314.

Quarstein, John V. "The Capture of Gosport Navy Yard." *Civil War Navy—The Magazine,* Vol. 5, Issue 4 (Spring 2018): 13-23.

Welles, Gideon. "Mr. Welles in Answer to Mr. Weed, The facts of the Abandonment of the Gosport Navy-Yard." *The Galaxy,* X (July 1870): 15-35.

Further Reading

Boynton, Charles B. *The History of the Navy During the Rebellion.* New York: D. Appleton and Company, 1867.

Campbell, R. Thomas and Alan B. Flanders. *Confederate Phoenix, The CSS Virginia.* Shippensburg, PA: Burd Street Press, 2001.

Dickon, Chris. *Chesapeake Bay Steamers.* Charleston: Arcadia Publishing, 2006.

Foss, William O. *The United States Navy in Hampton Roads.* Norfolk/Virginia Beach: The Donning Company/Publishers, 1984.

King, Captain Kendall, USN (Ret). "Disaster at Gosport." *Naval History,* Vol. 10, No. 2 (April 1996): 20-24.

Konstam, Angus. *Duel of the Ironclads.* Botley, Oxford: Osprey Publishing, Ltd., 2003.

Mapp, Alf J. and Ramona. *Portsmouth, A Pictorial History,* Expanded Second Edition. Virginia Beach, VA: The Donning Company Publishers, 1989.

Marolda, Edward J. *The Washington Navy Yard, An Illustrated History.* Washington: The Naval Historical Center, 1999.

Miller, Lt. Christopher, USN (Ret). *Norfolk Naval Shipyard.* Charleston: Arcadia Publishing, 2018.

Park, Carl D. *Ironclad Down, The USS Merrimack-CSS Virginia, from Construction to Destruction.* Annapolis: Naval Institute Press, 2007.

Selfridge, Captain Thomas O. Jr., USN. "The *Merrimac* and the *Cumberland.*" *The Cosmopolitan,* Vol. XV (May-October 1893): 176-84.

Museums, Libraries, Etc.

Fondren Library, Rice University, Houston, Texas

Hampton Roads Naval Museum, Norfolk, Virginia

Mariners' Museum, Newport News, Virginia

National Museum of the United States Navy, Washington, D.C.

Portsmouth Naval Shipyard Museum, Portsmouth, Virginia

U.S. Naval Academy Museum, Annapolis, Maryland

Websites

American Battlefield Trust, "The Decision at Craney Island," by Christopher Pieczynski, https://www.battlefields.org, accessed February 2022.

Encyclopedia Virginia:

"John Letcher (1813-1884)," https://encyclopediavirginia.org/entries/letcher-john-1813-1884/, accessed December 2022.

"Henry A. Wise (1806-1876)," https://encyclopediavirginia.org/entries/wise-henry-a-1806-1876/, accessed December 2022.

Find A Grave, "Como Charles Stewart McCauley," https://www.findagrave.com/memorial/37275570/charles-stewart-mccauley, accessed July 2021.

Milestones in the History of U.S. Foreign Relations, "French Intervention in Mexico and the American Civil War, 1862-1867," https://history.state.gov/milestones/1861-1865/French-intervention, accessed December 2022.

Naval History and Heritage Command (NHHC):

"Annual Reports of the Secretary of the Navy," https://www.history.navy.mil/content/history/nhhc/research/library/online-reading-room/title-list-alphabetically/a/secnav-reports.html, accessed January 2023.

"Dictionary of American Naval Fighting Ships (DANFS)," https://www.history.navy.mil/research/histories/ship-histories/danfs.html, accessed November 2022.

"Officers—Continental & U.S. Navy/Marine Corps 1775-1900," https://www.history.navy.mil/research/library/online-reading-room/title-list-alphabetically/o/officers-continental-usnavy-mc-1775-1900.html, accessed July 2022.

"Officers List—who resigned at the outbreak of the Civil War," https://www.history.navy.mil/research/library/manuscripts/o/officers-list-who-resigned-at-the-outbreak-of-the-civil-war.html, accessed August 2021.

Norfolk Navy Yard:

"History Matters," http://www.usgwarchives.net/va/portsmouth/shipyard/nnyblog.htm, accessed October 2022.

"Norfolk Navy Yard Table of Contents," http://www.usgwarchives.net/va/portsmouth/shipyard/nnytoc.html, accessed February 2022.

Virginia Places:

"Norfolk Naval Shipyard," http://www.virginiaplaces.org/military/norfolknavalshipyard.html, accessed July 2021.

"Before the Pentagon, Where Were the Military Headquarters?" http://www.virginiaplaces.org/military/beforethepentagon.html, accessed February 2019.

West Gulf Squadron, "General Naval Order, National Republican, Washington DC, 4 May 1861," www.westgulfsquadron.net/articles/generalorder.html, accessed December 2022.

* * *

INDEX

NOTE: Officers' ranks or diplomatic assignments described herein are those held during the period of this book's narrative and may not fully reflect that officer's ultimate rank or that diplomat's subsequent service.

A

Act for the Gradual Increase of the Navy, 5-6, 49
Agamemnon, H.M.S., 42
Alabama, C.S.S., 159, 221
Alden, James (commander, USN), coast survey service, 66; Gosport relief expedition, 155, 190, 193-94, 199; later service and death, 220; *Merrimack* delivery, 65-67, 73, 77, 79-81, 122, 126, 137-40, 191, 196; mission failure, 82, 140, 143-45, 169, 217; report to Welles, 147-49, 153, 157, 191, 206
American Revolution, 2
Anacostia, U.S.S., 73, 153
Anderson, Robert (major, USA), 23-24, 47, 70-71, 133
Annapolis, Maryland, 65
Archbold, Samuel (engineer-in-chief, USN), 61
Armstrong, James (captain, USN), commandant of Pensacola NY, 23-24, 26-27; court of enquiry, 27-29, 43-44, 213; court-martial, 44, 205-06, 211
Atlantic Hotel, 61, 79, 92, 145

B

Baldwin, John (VA delegate), 59, 89, 130; meeting with Lincoln, 59-60
Baker, Adam N. (first lieutenant, USMC), 53, 160
Baltic, S.S., 57, 133
Baltimore, Maryland, 1, 12, 77, 135, 151, 156, 177; riots, 150, 155, 203
Barron, James (commodore, USN), 3, 50
Barron, Samuel (captain, USN), 25, 50
Barron, Susan Virginia (later Susan Virginia Pendergrast), 50; *See also* Pendergrast, Susan Virginia.
Bay Line (Baltimore Steam Packet Company), 77-78, 86, 122, 135, 145, 147, 155-56, 170, 173, 177
Beauregard, Pierre G.T. (general, CSA), 70-71
Bennett, Frank M. (passed engineer, USN), correspondence with Isherwood, 61; *The Steam Navy of the United States*, 61; uncomfortable truth, 212-13
Black Warrior incident, 18
Blair, Mongomery (U.S. postmaster general), 46-47
Border states, 33, 38, 46, 59, 90

Boston, Massachusetts, 14, 220
Boston, U.S.S., 14, 16
Bradford, John (paymaster, USN), 214
Brazil Station, 14, 16, 20, 42
Breese, Samuel L. (captain, USN), 55, 125, 135, 155
Brooklyn, U.S.S., 25
Brooklyn (New York) Navy Yard, 12, 55-57, 65, 125, 135, 155, 197, 214
Brown, John, 35, 157
Brownley, James, 199
Buchanan, Franklin (captain USN, CSN), captain of C.S.S. *Virginia*, 208-09; commandant of Washington NY, 153
Buchanan, James (U.S. president), 22, 24-25, 47, 215-16
Bureau of Construction, Equipment and Repairs, 41

C

Calendar, annotated, Appendix 1
Cameron, Simon (U.S. secretary of war), 74
Cassin, John (commodore, USN), 4-5
Charleston, South Carolina, 21, 47, 57, 63, 70
Charlestown (Boston) Navy Yard, 50, 151
Chase, Salmon (U.S. treasury secretary), 37
Chauncey, Isaac (commodore, USN), 12
Chesapeake, U.S.S., 3, 50, 56, 179
Chesapeake Bay, 1, 77-78, 145, 147, 151, 177
Civil War, 68-69, 142, 148, 201, 203, 210, 212
Clements, James, 165
Cockburn, George (admiral, RN), 5
Columbia, U.S.S., 8, 189, 201; burning, 194, 200
Columbus, U.S.S., 7, 189; proposed conversion, 201, Appendix 6; scuttling, 174, 200
Confederate States of America, founding, 29-31; Virginia's admission, 206
Congress, U.S.S., 208-09
Constellation, U.S.S., 4-5, 12-13, 18
Constitution of the U.S., 33
Constitution, U.S.S., 4, 73
Crab Creek, 2
Crane, William (captain, USN), 14
Craney Island, 1, 4, 79, 92, 122, 158, 194; Battle of, 5, 12, 20, 204
Crawford, William H. (U.S. minister to France, secretary of treasury), 15
Crittenden, John (U.S. senator), 32; compromise, 32, 37-38, 59
Cuba, 18
Cumberland, U.S.S., 79, 201; arrives at Hampton Roads, 49; description, 49, 79, 169; escape from Gosport, 192, 195-97, 203; fight with *Virginia*, 208-09, 221; Gosport guardship, 79, 85, 92, 119-20, 135, 158, 161, 163-64, 167-68, 170-71, 173, 175, 177-78, 180, 182-83, 185, 187, 191-92; relocated nearer yard, 121-23, 125; service under Pendergrast, 54, 73, 85, 214
Cyane, U.S.S., 52

D

Dahlgren guns, 42, 49, 187, 208
Dahlgren, John A. (commander, USN), 17, 42, 135, 153
Danby, Robert (chief engineer, USN), 54, 61-62, 64, 80, 126, 137
Davis, Jefferson (C.S. president), 29, 157

Decatur, Stephen (captain, USN), 11-12
Delafield, Richard (colonel, USA), 151, 153-54
Delaware, U.S.S., 6, 8, 16-17, 189, 201; scuttling, 174, 200
Democratic party, 21
Dickins, Asbury, 14-16
Dickins, Lilia Elzabeth (later Lilia Elizabeth McCauley), early life, 14; marriage, 14; *See also* McCauley, Lilia Elizabeth.
Dimick, Justin (colonel, USA), 177, 195
Dolphin, U.S.S., 7, 28, 161, 171; burning, 194, 200; removal from Gosport, 84-85, 123, 155; scuttling, 173-74, 180; temporary guard ship service, 119, 122-23, 158-59, 175, 221
Donaldson, Edward (lieutenant, USN), 53, 144, 160, 178, 205
Doran, Edward (paymaster, USN), 145, 147, 159, 214
Du Pont, Samuel Francis (captain, USN), 25, 50, 135, 142, 155, 195, 214, 219

E

Edelin, James (lieutenant colonel, USMC), 53, 160, 179, 196, 205, 214
Efficiency Board, 18-19
Elizabeth River, Virginia, 1-2, 4-5, 7, 51, 121, 144, 161, 169, 184, 192, 209; channel obstructions, 79, 90-93, 129, 131, 138-39, 143-44, 150, 158, 177, 181, 195, 217-18; eastern branch, 35, 161; southern branch, 2, 161, 172, 175, 189, 193, 200
Engineer Corps, 62
Erie, U.S.S., 13, 16
Exchange Hotel, 91, 127

F

Fairfield, U.S.S., 15
Falmouth, U.S.S., 15
Florida, secession, 26, 74
Fort Barrancas, 24, 26
Fort Norfolk, 6, 131, 158; seizure of arsenal, 163-64, 167, 169-70, 182, 201
Fort Pickens, 24, 26-27, 40, 47, 57, 69-70, 85, 125, 204, 217
Fort Sumter, 23-25, 40, 85, 204; hostilities against, 69-71, 74-75, 79, 84, 175, 203; relief expedition, 47, 54, 56-57, 60, 63, 65, 133, 153, 167, 217
Fortress Monroe, 2, 31, 37, 77, 148, 151, 155-56, 169, 173, 177, 180, 182, 195-97, 209
Fox, Gustavus Vasa, 47, 56, 63, 69-71, 133, 153, 167
Fox, U.S.S. 14
Frigates, 2, 41, 49, 201

G

Germain, Captain Charles, 133, 197
Georgeanna, P.S., 78
Germantown, U.S.S., 7, 84-85, 171, 174; burning, 193-94, 200; damaged by sheers, 174-75, 182, 200; preparing for sea, 28, 45, 55, 58, 65, 155; scuttling, 173-75
Gosport (Norfolk) Navy Yard, 31, 35, 37, 68, 77-78, 127, 129, 133, 148, 151, 154, 156, 159-65, 169, 172-74, 181, 203, 221; command structure, 8, 52, 139; defenses, 28, 46, 94, 150, 158, 160-61, 167, 170-72, 181-83; description, 6, 68, 80, 162; destruction and abandonment, 175, 193-96, 199, 203, 209-10, 212, 216; early years, 2-7, 12; federal purchase, 3; front gate, 121, 161, 181, 190; graving dock, 6, 149, 174, 187-89, 193-94, 199,

207-08, 210; gun depot, 131; guns seized by South, 201, 207, Appendix 5; guns spiked or damaged, 163, 170, 173, 182, 187; *Merrimack*'s lay-up, 50, 64; occupation by Virginia, 199; officers, 9, 73-74, 120, Appendix 2; origins, 2-3; prominence among navy yards, 1, 6, 173, 204, 211; public reaction to loss, 202-03; scuttling of ships, 172-73, 178, 180, 182; security concerns, 27-28, 55, 158, 162; sheers, 173-74, 182; ship houses, 7, 121, 190, 194, 200; Southern officers, 9, 54, 73, 75, 120-22, 131, 138, 171, 205, 216, Appendix 2; Southern officer resignations, 139, 143, 149, 159, 170, 179, Appendix 2; stores and workshops, 131, 149, 163, 190, 200-01, 207; value of lost property, 201-02, 211; warships at yard, 7-8, 16, 90, 120, 158, 170-71, 189-90, Appendix 4; warship potential for conversions, 42, Appendix 6; worker disloyalty, 25, 50, 80
Greenleaf Point, 156
Gregory, Francis H. (captain, USN), 20, 121, 214
Grimes, James (U.S. senator), 212

H

Hale, John P. (U.S. senator), 212, 215
Hampton Roads, Battle of, 169, 207, 209-10, 212; waters of, 1-2, 5, 12, 16, 20, 49, 151, 192, 194, 204, 208-09, 212, 214
Harper, Kenton (general, VA militia), 127, 129, 147, 157
Harpers Ferry, Virginia, arsenal, 31, 151; seizure, 31, 35, 37, 68, 90-91, 127-28, 147-50, 157, 203, 216
Harriet Lane, U.S.S., 54-57, 73, 171
Harris, John (colonel, USMC), 154
Heth, Henry (colonel, VA militia), 157, 165
Home Squadron, 18, 20, 42, 49, 56, 73
Hunter, William W. (commander, USN), 135, 155

I

Imboden, John D. (captain, VA militia), 91, 93, 127-29
Irwin, John (lieutenant, USN), 73-74, 135, 169, 173
Isherwood, Benjamin F. (engineer-in-chief, USN), background, 61-62; channel blockage, 92, 137; later service and death, 220-21; mission failure, 141-42, 145, 169; report to Welles, 62, 147-49, 157; restoring *Merrimack*'s machinery, 62, 67, 73, 77-80, 82, 84, 90, 122, 126, 137-38, 216-17; Senate testimony, 92, 215

J

James River, 1, 81, 182
Jefferson, U.S.S., 12
Jones, Catesby ap Roger (lieutenant, USN, VN, CSN), command in Norfolk, 157; *Virginia* service, 209
Junon, H.M.S., 5

K

Keystone State, S.S., 196-97, 211; Paulding shifts his flag, 195; preparation in Philadelphia, 136
King, James (chief engineer, USN), 188
Kintzing, Matthew (captain, USMC), 163

L

Leopard, H.M.S., 3
Lenthall, John (chief, Bureau of Construction, Equipment and Repairs), 19, 61-62, 64, 72
Letcher, John (Virginia governor), 33, 37, 90-91, 128-29, 157
Light-boats, scuttling, 92-93, 131
Lincoln, Abraham (U.S. president), 134-35, 148-49, 151, 153, 215-16, 218; administration policies, 60, 63, 69-72, 74, 167, 204, 217; assumed office, 29; cabinet, 40, 46, 148-49, 153; call-up of militia, 72-74, 84, 89-90, 203; election, 21-22; inauguration, 38-40; martial orders, 47, 54, 56; proclamation of blockade, 150, 155, 203; Virginia policies, 44-45, 71, 83
Livingston, John (commander, USN), 73, 160, 169, 221-22
Loyall, George (navy agent), 53

M

Macedonian, U.S.S., 23, 73
Mahone, William (general, CSA), 161-62, 182
Mallory, Stephen R. (U.S. senator, C.S. secretary of the navy), 207; quest for ironclad, 207-08
Marston, John (captain, USN), 49, 119, 138-39, 195, 218
Massachusetts Volunteers, 150-51, 177, 185, 191-92, 195
McCauley, Charles Stewart (captain, USN), Battle of Craney Island, 5, 12, 144, 204; board service, 18-19; captain service, 16-20; commandant of Gosport NY, 1, 7, 19, 25, 28, 45, 50, 55, 58, 64, 68, 80, 82-86, 93-94, 119-22, 134, 142-45, 152, 155-56, 159-61, 164-65, 173, 175, 178, 191, 197, 205-06, 211-12, 218, 222; commandant of Washington NY, 1, 17; death, 220; denied needed crews, 57-58, 120, 125, 135, 170, 215; early life, 11-12; failure to request court of enquiry, 213; home on F Street, 19, 212, 214; home squadron service, 18; later service, 212, 214, 220; lieutenant service, 12-13; master commandant service, 14-16; merchant marine service, 13, 20; *Merrimack* decision, 126, 137-38, 140-45, 147-48, 217; midshipman service, 4-5, 11, 204; orders to move ships, 50, 55, 64, 84-85; overwhelmed by events, 141-43, 148, 179, 213; relieved of duty, 179, 212; report to Welles, 82, 173; retirement, 220; scuttling of ships, 172-76, 178, 181, 184, 204, 217; Senate censure, 216; Senate testimony, 82, 141, 144, 162, 165, 168, 170, 172, 214-16, 220; yard defense worries, 158, 162, 164, 170-72
McCauley children, 15-20
McCauley, Daniel Smith (midshipman, USN), 13-14, 19
McCauley, Edward, 18, 141; tearful plea, 191
McCauley, Lilia Elizabeth, 15-16, 19; *See also* Dickins, Lilia Elizabeth.
McGary, Charles (lieutenant, USN), 189
Mediterranean Station, 13, 17, 20, 42
Meigs, Montgomery (captain, USA), 47, 69-70
Mercer, Samuel (captain, USN), 56, 63, 69
Merrimack, U.S.S., at Gosport, 78-80, 171, 201; burning, 189, 193-94, 210, 200; crew and officers, 53, 65-66, 82, 85, 155;

decommissioning, 7; description, 42, 81, 196, 215-16; grounding in 1856, 144; ironclad conversion, 207; machinery, 7, 62, 64, 67, 77, 80, 137-38, 173, 207, 216; removal from Gosport, 80-82, 93, 120, 122-23, 126, 137, 143-44, 155, 170, 176, 183, 185, 191, 196, 206-07, 210, 217; scuttling, 173-74, 180, 200, 207; Welles' orders, 55, 58, 61-62, 65-67, Appendix 3
Mexican War, 1, 157
Minnesota, U.S.S., 208, 210
Mitchell, Captain, 177
Monitor, U.S.S., 210, 214
Montague, Robert L. (VA delegate), 129
Montgomery, Alabama, 29
Morgan, Charles (commodore, USN), 16
Morris, Charles (commodore, USN), 16
Morris, George (lieutenant, USN), 189
Murphey, Peter U. (lieutenant, USN), 53
Murray, Alexander (lieutenant, USN), 138

N

Navy Department, 41, 56, 61, 69, 79, 120, 148, 205, 212, 214; bureaus, 41, 221; morale, 43
Navy Register, 20, 41, 73
Newport News, 81
New York, U.S.S., 7, 190, 200
New York City, 57, 133, 203
Nicholson, A. S. (major, USMC), 214-15
Niven, John, *American Secretaries of the Navy*, 57, 204
Norfolk & Petersburg Railroad, 161-62
Norfolk Navy Yard, 31, 45, 50, 67, 87, 90, 128, 135, 149, 151, 153, 155, 157, 219-20, 221-22; *See also* Gosport Navy Yard.
Norfolk, Virginia, city of, 1-7, 35, 51, 77-80, 91-93, 133, 145, 150-51, 159, 161, 164, 167, 169-70, 177, 182, 191, 194, 204, 208; vigilance committee (safety, etc.), 92-93, 133, 157; volunteer companies, 93, 158, 163, 182
North, James (lieutenant, USN, CSN), 207

O

O'Connor, Daniel (private, USMC), 168-69
Office of Detail, 43, 83, 154, 211, 213
Official Records of the Union and Confederate Navies in the War of the Rebellion (Official Records), 28-29, 45, 55, 62, 64, 66, 119, 183, 200, 205, 213
Old Dominion, *see* Virginia.
Old Navy Department building, 19, 41
Old Point Comfort, 1-2, 31, 69
Oliver, Charles B. (gunner, USN), 9, 160
Ontario, U.S.S., 12
Ordinance of Secession, Virginia, 60, 91, 130, 134, 139, 157, 159, 175, 199; Ordinance No. 1, 131-32; Ordinance No. 9, 131-32; ratification, 132
Ordinary, 7-8, 42, 53, Appendix 4
Ordnance, 52, 131, 158

P

Pacific Station, 13, 15, 18, 20, 42
Page, Richard L. (commander, USN, CSN), 9, 53, 135; later service, 221; plea for Southern officers, 122-23, 133, 160; relative of Robert E. Lee, 53
Parker, William H., 91

Parrott, Enoch (lieutenant, USN), 73
Paulding, Hiram (flag officer, USN), decision to destroy Gosport, 172, 181-87, 192-93, 204, 212-13, 218; early career, 83; flying visit to Gosport, 90, 119-25, 134-35, 170, 218; Gen. Walker confrontation, 83; Gosport relief expedition, 151-56, 173, 177-83, 187, 189, 190-92, 194-97, 205-06, 211, 217; Home squadron service, 83; information on secessionists, 182-83; later service and death, 211, 213-14, 220; report to Welles, 134, 149, 153, 181, 205; Senate testimony, 124, 181-83, 214; use of force, 184-85; Welles' orders to, 85-86, 152-53
Paullin, Charles, *History of Naval Administration,* 204
Pawnee, U.S.S., 54, 56, 73, 218; description, 154; escape from Gosport, 194-97, 203, 205, 211; Gosport service, 153-56, 177-80, 182-85, 187-89, 191-92
Peace Conference, 31-32, 37-38
Pegram, Robert (lieutenant, USN; captain, VN, CSN), 157-58, 164
Pendergrast, Garrett J. (captain, USN), background, 50; flag officer of Home Squadron, 43, 49-51, 84-86, 93, 119, 121-22, 134-35, 138, 142, 144, 161, 164, 175, 187, 195, 197, 211-12, 218; later service and death, 214, 219-20; marriage to Susan Virginia Barron, 50, 218; *Merrimack*'s captain, 50, 144, 218; Senate testimony, 52, 167-68, 171-72, 179, 214; views on flogging, 50; *Yankee* potential, 168-69
Pendergrast, Susan Virginia, 50, 93, 220; *See also* Barron, Susan Virginia.
Pennsylvania, U.S.S., 187, building, 6, 16; burning, 193-94, 200; receiving ship, 6-8, 16, 53, 73, 121, 158, 161, 170-71, 173, 175, 178, 182, 189, 192, 214
Pensacola, U.S.S., 53
Pensacola Navy Yard, 23-24, 42; seizure, 26-28, 74, 170, 211, 213, 216, 219
Perry, Matthew (captain, USN), 19
Perry, U.S.S., 73
Peters, William (paymaster, VN, CSN), 174, 199
Phelps, Ledyard (lieutenant, USN), 189, 192
Philadelphia, Pennsylvania, 11, 220
Philadelphia Navy Yard, 6, 16, 67, 125, 135, 151, 195, 214, 219
Phillips, Dinwiddie B. (passed assistant surgeon, USN), 9
Pierce, Franklin (U.S. president), 18
Plymouth, U.S.S., 7, 84-85, 171; burning, 189, 200, 194; preparing for sea, 28, 55, 58, 65, 123, 155; scuttling, 173-74, 200
Pocahontas, U.S.S., 54, 56, 73
Poindexter, Carter B. (lieutenant, USN), 9, 53, 160
Pook, Samuel (naval constructor), 54
Porter, David D. (lieutenant, USN), 69-70, 141, 168, 201-02, 220
Porter, John L. (naval constructor), 9, 54, 125
Portsmouth (Kittery) Navy Yard, 23
Portsmouth, Virginia, town of, 1-7, 50-51, 77, 80, 167, 174, 177, 188, 194; volunteer companies, 93, 158, 182, 199
Portsmouth Naval Hospital, Portsmouth, Virginia, 6

Potomac River, 1, 32, 56, 150, 153, 154, 156
Powell, Enoch (lieutenant, USN), 169
Powell, Levin (captain, USN), 27-28, 45-46, 55, 83, 87
Powhatan, U.S.S., 55-57, 63, 69-70, 73
Preston, William B. (VA delegate), 60, 127, 130

Q

Quasi-War, 3

R

Randolph, George Wythe (VA delegate), 33, 60, 67-68, 130
Raritan, U.S.S., 18, 189, 201; burning, 194, 200
Razee, razeed, 49
Republican party, 21, 37-38, 46
Reynolds, John (ordinary seaman, USN), 189, 194
Rich, Jabez (captain, USMC), 53, 160
Richmond, Virginia, 1, 38, 71, 89-91, 169, 221
Ridgely, Charles (commodore, USN), 13
Roanoke, U.S.S., 208
Robb, Robert G (commander, USN), 52, 68, 73, 82, 135, 139, 160
Rodgers, John (commander, USN), 155, 187-89, 193-95
Ronckendorff, William (lieutenant, USN), 125
Rootes, Thomas (commander, USN), 53, 135
Rowan, Stephen (commander, USN), 154
Royal Navy, 2-5
Russell, John (lieutenant USN), 188, 194, 214
Russell, William Howard, correspondent, 78-79; *My Diary, North and South,* 78

S

Sabine, U.S.S., 23
Sackets Harbor, 12
Saint Helena annex, 6, 161-62; battery earthworks, 162, 164-65, 167, 169-71, 181
St. Lawrence, U.S.S., 83, 208
St. Louis, U.S.S., 15
Sands, Benjamin (commander, USN), 190, 193-94, 199
Savannah, U.S.S., 18
Scharf, Thomas, 92
Scott, Robert (VA delegate), 75-76, 127
Scott, Winfield (general, USA), 46, 150-53, 177
Scuttling ships, 172
Seamen's protection certificate, 11
Secession, Southern, 21-23, 29, 31, 39, 145, 150, 163, 203
Selfridge, Thomas O. Jr. (lieutenant, USN), Battle of Hampton Roads, 209-10; *Cumberland* service, 119, 121-23, 164-65, 173, 177, 195, 209-10; later service and death, 221; views on Gosport destruction, 184-85
Semmes, Alexander A. (lieutenant, USN), 52, 159, 161, 205
Semmes, Raphael (captain, CSN), 52, 159
Senate Committee on Naval Affairs, 212, 214
Senate select committee, core findings, 216-18; judgements, 124-25, 219; origins, 212; report (No. 37), 162, 202, 216, 219; testimony, 52, 120-21, 124, 213-16
Sewell's Point, 92, 158, 161, 169, 195
Seward, William (U.S. secretary of state), 46-47, 54, 69-70
Sharp, William (lieutenant, USN), 52-53

Shubrick, William (captain, USN), 19
Sinclair, Arthur (commander, USN), 9, 52-53, 135, 160
Sinclair, George T. "Terry" (lieutenant, USN, VN, CSN,), 52, 73, 139, 160, 199, 221; seizure of arsenal, 163
Slemmer, Adam (first lieutenant, USA), 24, 26
Sloops-of-war, 41-42, 201
South Carolina secession, 21, 59
Spiking guns, 163
Spotswood, Charles F.M. (lieutenant, USN, VN, CSN), 52, 73, 135, 139, 199
Sprowle, Andrew (navy agent), 2
Star of the West, 25
Steam frigates, 42, 207
Stevens battery, 34, 41
Stewart, Charles (captain, USN), 1, 4, 12, 16, 19, 220
Stringham, Silas (captain, USN), 43
Storer, George (captain, USN), 121
Stuart, Alexander H.H. (VA delegate), 60
Summers, George (VA delegate), 129
Supply, U.S.S., 26

T

Taliaferro, William B. (general, VA militia), 35, 157-59, 162, 164-65, 180-82, 191
Tarbell, Joseph (master commandant, USN), 4-5
Toucey, Isaac (U.S. secretary of the navy), 23, 27-28, 43, 49, 52, 83, 200-01, 215
Tredegar Iron Works, 208
Truxton, Thomas (commodore), 3
Tucker, John R., (commander, USN), 9, 52, 82, 135, 138-39, 160, 221; Norfolk ties, 9, 52
Tyler, John (U.S. president), 31; VA delegate, 129
Tyler, Nathaniel (major, VA militia), 157

U

United States, U.S.S., 8, 12-13, 189, 200
U.S. Army, strength, 22, 72
U.S. Naval Academy, Annapolis, Maryland, 55, 65
U.S. Navy, cruising stations, 20, 42, 72; frigates, 2, 41; officer seniority, 86, 121, 124, 149; strength, 22, 41-42, 72, 150

V

Vandalia, U.S.S., 53
Vera Cruz, Mexico, 42, 49, 54, 85
Virginia, Commonwealth of, 2-3, 31-32, 35, 72, 74, 150, 157, 199, 207; provisional navy, 164, 182; secession, 32, 59, 76, 89, 130-34, 136, 139-40, 203-04, 217
Virginia secession issues, 32-33, 132
Virginia State Convention, 31-33, 38, 59-61, 67-68, 71, 74-75, 80, 91, 120, 122; resolutions, 60-61, 127, 129-30; secret session, 76, 89-90, 127-33
Virginia, C.S.S, Battle of Hampton Roads, 209-10; description, 208-09; *Merrimack* origins, 126, 169, 207-08, 210, 215, 221
Virginian naval officers, 131-33

Y

Yankee, steam tug, arrival in Norfolk, 133; attempted seizure, 167, 170-71; chartered by Fox, 70, 133, 167, 182; description, 133, 168; potential as gunboat, 168-69, 173, 182, 218; towing *Cumberland*, 169, 192, 195-97, 203

Yeo, Sir James (captain, RN), 12

W

Wabash, U.S.S., 53, 83
Walker, William (filibuster general), 83
War Department building, 150
War of 1812, 2, 4-5, 12, 50, 52, 79, 188, 204, 210
Ward, James H. (commander, USN), 184
Wardrop, David W. (colonel, Massachusetts Volunteers), 177, 187
Washington, District of Columbia, 1, 5, 31, 69, 122, 125, 150-51, 157, 196-97, 203
Washington, George (U.S. president), 31
Washington Navy Yard, 12, 17, 83, 135, 153-54, 208, 211
Water Witch, U.S.S., 73, 125
Watters, John (lieutenant, USN), 53, 73
Welles, Gideon (U.S. secretary of the navy), 68-69, 73, 151, 153-54, 157, 197, 211; annual report, 218-19; appointment, 40; Armstrong court-martial general order, 205-06, 211, 213; assumed office, 28, 40, 42-43; blunders and bungling, 57, 70-71, 152, 155-56, 173, 204, 217; death, 220; diary, 55, 68, 70, 73, 134, 140, 148, 175-76; early life, 40-41; judgements of McCauley and Pendergrast, 175-76; orders to Alden, 65-66, 138, 140; orders to McCauley, 45, 55, 62-65, 67, 80, 84, 86, 120, 123-25, 144, Appendix 3; orders to Paulding, 83-87, 90, 124, 134, 152, 184; orders to Pendergrast, 51, 54; 84-86, 123-24, Appendix 3; orders to replace Southern officers, 73-74, 140; orders for vessel and crew mobilizations, 56-58, 65, 69, 85, 125, 136, 154, 216; political views, leadership, 41, 46, 217, 220; Senate committee interactions, 212, 215
West Indies Station, 15, 20, 42
White House (Executive Mansion): 41, 148, 150
Wilkes, Charles (captain, USN), Gosport relief expedition, 141, 154, 178-79, 187, 191, 193-94, 218; opinion of Paulding, 84; report to Paulding, 178, 205; views on yard's destruction, 183-84, 186, 205
Willard's Hotel, 32, 68
Winder Building, 19, 41
Wise, Barton, 34, 75
Wise, Henry A. (ex-Virginia governor, general, CSA), 157; appearance, 34; beliefs, 36-37, 59, 89, 129-30; convention delegate, 34, 60-61, 74-76, 89-90, 127-31; early life, 34-35; later life and death, 39, 221; political life, 33-34; private insurrection (aggression towards Harpers Ferry and Gosport), 91-92, 127-29, 147, 157
Wise, Henry A. (lieutenant, USN), 155, 180, 184, 188-89, 192, 194, Senate testimony, 193, 214
Wood, Fernando (N.Y. politician), 39
Woodhull, Maxwell (lieutenant, USN), 136, 195
Wright, Horatio G. (captain, USA), 154, 177-78; mining dock, 187-89, 194; report, 179, 188, 206
Wyandotte, U.S.S., 26, 7

* * *

www.ingramcontent.com/pod-product-compliance
Lightning Source LLC
Chambersburg PA
CBHW062028090726
47698CB00024BA/345/J
* 9 7 8 1 6 4 8 8 9 8 0 8 2 *